THE PROJECT SURGEON

THE PROJECT SURGEON

A Troubleshooter's Guide to Business Crisis Management

Boris Hornjak

Project Management Institute

Library of Congress Cataloging-in-Publication Data

Hornjak, Boris, 1961–
The project surgeon : a troubleshooter's guide to business crisis management / Boris Hornjak.
p. cm.
Includes index.
ISBN: 1-880410-75-3 (alk. paper)
1. Crisis management. 2. Emergency management. I. Title.

HD49 .H67 2001
658.4'056 – – dc21 2001016017
CIP

ISBN: 1-880410-75-3

Published by: Project Management Institute, Inc.
Four Campus Boulevard
Newtown Square, Pennsylvania 19073-3299 USA
Phone +610-356-4600 or visit our website: www.pmi.org

The paper used in this book complies with the Permanent Paper Standard issued by the National Information Standards Organization (Z39.48—1984).

10 9 8 7 6 5 4 3 2 1

Contents

In this book, we saw no satisfactory workaround to using *EV* to abbreviate *expected value*. We hope this does not cause undue confusion with the *earned value* concept in project management.

List of Figures

List of Tables

Introduction

No problem can stand the assault of sustained thinking.

Voltaire

The intent of this book is to provide a best practice primer for operational crisis management, as well as for company and project turnarounds in a general business environment. It is written as a hands-on troubleshooting manual for operational managers, the corporate combat soldiers fighting daily battles in the trenches of business warfare.

There is nothing in this book for those who deal in the aftermath of a crisis—the arbitrators, media handlers, litigators, or public relations specialists. For these and other assorted blame experts, there is a plethora of books dealing in the exculpation and explanation of failure.

The focus of this book is on business recovery, not on dealing with failure—you are the business emergency room physician, a MASH surgeon, and an intensive care nurse, not a forensic pathologist or a coroner. This book is written from the operator's standpoint for a practical manager thrust into a crisis situation with a mission to turn things around, make tough decisions under fire, address problems when they occur, and prevent them from happening again. Its intent is to arm you with practical tools and methods to correct the immediate symptoms of a crisis situation, address its underlying causes, and build crisis-prevention mechanisms.

I understand that you cannot carry a reference library to address each and every crisis situation that you may encounter, nor do you have the time to look up the solutions. Therefore, I have condensed the essential information in a bullet-point/checklist/flow-chart format for consulting under fire in the corporate battlefields. Major points throughout the book are coded: arrows for definitions, checkmarks for action lists, and square bullets for summaries.

If this book is written for you, you probably have barely enough time to read this far. This in itself should be a compelling reason to continue.

The Value of Crisis Management

> The best money is to be made when there's blood in the streets.
>
> J. P. Morgan

Effective crisis management can mean the difference between survival and failure for a troubled project, or even an entire company. However, the objective of crisis management is always to minimize losses, not to maximize gains—the latter objective is accomplished with effective crisis prevention, which will be discussed later in the book.

There is nothing glamorous about troubleshooting; to paraphrase management guru Peter Drucker, it amounts to feeding the problems and starving the opportunities. The best people in companies are routinely assigned to "problem solving" instead of creating new inventions, business opportunities, or growth. They are minimizing losses and containing damage instead of maximizing gains. The opportunities have to fend for themselves. New projects are assigned to substandard personnel because the stars are busy troubleshooting old projects. Or, the top performers are assigned to start up a new project, but are then pulled away for firefighting duty. By the time the new projects are mismanaged into a crisis situation, the troubleshooters are called in, and the whole cycle is repeated.

Why is this the case? Because on the corporate value scales, the potential negative consequences of failure more than offset the potential positive gains of success. It is much easier to slip and fail than to try hard and succeed.

Great accomplishments and inventions were never achieved as a result of bailing out of dire straits—the Apollo 13 mission, celebrated in a 1990s movie, was called a "success," although it failed to achieve its primary objective of landing on the moon and returning safely to Earth. Not only did it fail to land on the moon, but also its crew was lucky enough to beat almost insurmountable odds and return home alive. It was, however, a magnificent triumph of emergency management, improvisation, creativity, calmness under stress, and focused competence that can provide numerous examples for crisis management in the business world. The objectives change in a crisis situation, from the most ambitious to the basics, if only because salvaging the basics gives us a second chance to live to fight another day. The path to recovery and success is long and arduous. Great comeback stories are rare and thus more endearing.

So what are crisis management skills worth? Everything, if you have everything to lose. Their value is in correcting the errors and ensuring the basic survival of an enterprise, not in achieving what could have been, but never will be. If your company is faced with a $10 million loss that can be reduced to a $2 million loss by spending $500,000 on crisis management, it sounds like a pretty good deal. But that's just the first step. The real value is added when you spend another $500,000 for a crisis prevention program to ensure that the $10 million mistakes are never made again. Consequently, you can free your best and brightest people to focus on opportunities instead of on problems, and ultimately break the failure/recovery cycle.

This book will touch upon a myriad of subjects, ranging from management and finance to engineering, statistics, and logic, but it is deliberately concise and action oriented and does not focus on any reference subject in depth. You, as the Project Surgeon, cannot be an expert in any one narrow technical field. Instead, you have to learn to draw upon and synthesize the specialist knowledge and resources while remaining a generalist: you must know what you don't know, know who knows, and put it all together quickly and effectively in a time of crisis.

Current Crisis Management Approaches

This book has evolved from a set of notes that I have compiled over the years of troubleshooting projects in the construction industry. It has evolved out of necessity, simply because I could not find a book that would give me practical tools to address various project crises when and where they occur. Every text that I consulted and everyone in the business to whom I spoke told me how to address the aftermath of the crisis situation and pass the buck.

The Project Surgeon includes a number of crisis management case studies and examples based on real experiences, both my own and those of my colleagues and coworkers. The worst crises make for the best case studies. Although based on real experiences, none of the cases and examples depicts actual projects, people, or events. To maximize their practical usefulness, and relevance to the "theoretical" portions of the text, I have thoroughly mixed the real-life project experiences among the case studies, resulting in fictionalized accounts of real-life situations. Thus, in case studies, the protagonist is simply called the "Project Surgeon."

Is there a career in crisis management, in being a business relief pitcher? Sure, if you approach it in a systematic, organized, and somewhat detached manner. Putting out fires is hard, precisely because it is repetitive—you reinvent the wheel every time the flames flare. You are so busy playing fireman that you have no time to devote to fire prevention and protection. Absence of a systematic fire-prevention approach leads to burnout. This book will provide you with a logical, systematic framework for addressing and preventing crises once and for all, eliminating the need to deal with the same type of problem over and over again.

I suppose that the reason so few people want to be business firemen is that it is easier to deal with the aftermath of failure; let the crisis run its course, cover your behind, pin the blame on someone else, let the "undertakers" of the business world—claims experts, lawyers, surety underwriters, expert witnesses, and others—pick up the pieces while you move on to other endeavors.

Before developing a new resolution-oriented strategy, let us briefly identify the four basic approaches to crisis management in today's business practice. These are more the informal ways that a company "culture" deals with problems, or reacts to them, rather than an organized and formal management process. What is common to all four is that they focus on the aftermath of a crisis, on the exculpation of failure, and on the protection of the culprits.

Belt-and-Suspenders Approach

As the name implies, the adherents of this approach believe that buying sufficient insurance will immunize them against problems. Figuratively speaking, they buy flood insurance and continue to live in the flood plains, but do not invest in protecting their homes from the effects of flood. Paradoxically, this risk-insurance approach is also known as *risk management*, although the risk is not managed prior to or when the exposure to the risk occurs. There is nothing wrong with being prudent in business, but this approach alone deals only with the negative aftermath of a crisis. It has to be coupled with proactive crisis prevention and management. It's like an obese chain-smoker with a $10 million life insurance policy. You are worth more dead than alive.

In real-life business, heavy dependence on bureaucratic procedures, lawyers, contractual language, responsibility shifting, and red tape and very little emphasis on the health of company fundamentals characterize this approach.

To summarize, the main points of the *belt-and-suspenders* approach are:

- Insurance against risk, rather than minimizing risk exposure.
- "Acceptable Losses" attitude.
- Illusion that it's easier and statistically cheaper to buy insurance against a loss than to correct it.
- Probability of a crisis occurring is so low that it does not warrant a proactive approach.
- Betting on the odds—the hundred-year flood won't happen in my lifetime, and if it does, it's not my fault.

Pin-the-Blame Approach

Practitioners of this approach are similar to the belt-and-suspenders people in their unwillingness or inability to deal with problems. However, they also recognize that their sloth will propagate and exacerbate the problems, so that they concoct often elaborate getaway plans to cover their tracks. Failure can always be blamed on someone else directly or indirectly involved in the crisis, while our involvement can be denied or minimized. If we do nothing, we can't do anything wrong. If things go well, we will swoop in and take credit; if they go wrong, we can deny involvement or even produce a strategic "I told you so" memo.

The plausible denial culture is prevalent in the heavily bureaucratic companies, where proper procedures are emphasized over results. In companies cultivating this approach, look for intensive office politics, backstabbing, brownnosing, responsibility avoidance, elaborate cover-up schemes and alibis, and heavy use of spin-doctors and PR consultants. Usually the effort expended in responsibility avoidance exceeds the effort that would have been required to do the job right in the first place.

In summary, this approach focuses on:

- Issue avoidance.
- "Teflonization" of key players.
- Alibis and plausible denials.
- Image preservation, damage control, denials, excuses, media control—spin doctors.
- Claims, litigation, arbitration.

Tombstone Approach

> I'm trying to figure out what we did wrong—that's what's troubling me. This company does not belong in bankruptcy.
>
> Jack Agresti, CEO of Guy F. Atkinson, upon declaring Chapter 11

This is the most indolent of all reactive approaches, characterized by a total disregard for the potentially disastrous consequences of inaction. Symptoms and warning signs of a crisis are ignored, not willfully but rather as a result of sloth and procrastination.

Symptoms do not warrant a preventive action, until disaster strikes. Measures are taken to correct a problem only after a catastrophe occurs, and then reluctantly and on a limited basis. Everyone carries a cynical "expected casualties" attitude. This is the ultimate "lazy company's" approach to crisis management.

Slash-and-Burn Approach

> You and you stay. The rest of you are fired. Good-bye.
>
> "Chainsaw" Al Dunlap, to the executives of a newly taken-over company

This approach cannot be strictly characterized as crisis management since it involves outsiders waiting in the wings to "turn around" or dismember a terminally ill company, rather than company insiders trying to stave off a disaster. Here are some characteristics:

- Five-to-midnight approach favored by corporate raiders and turnaround artists who come to the battlefield after the war and bayonet the wounded.
- This method has some excellent practitioners, and in many desperate cases, only radical surgery can save the continuity of the enterprise.
- Bottom fisher's approach—waiting in the wings until the target is ripe for the picking.
- The "saved" corporations emerge "leaner and meaner"—read "smaller and weaker"—with their stock value artificially enhanced due to reduced "costs." Companies are often acquired and sold for parts at a corporate chop-shop fire sale.

This method adds no economic value; it is the opposite of synergy, if separate parts are worth more than the whole. It treats assets as costs to be slashed; people, goodwill, and technology are sacrificed for short-term gain.

Conclusion

It is crucial for the Project Surgeon to be able to identify the tell-tale signs of the crisis avoidance phenomena in real life; more often than not you will be thrust into a situation where one or more of the previously described situations are taking place.

None of the above "methods" address the immediate symptoms and underlying causes of a crisis situation when and where the crisis occurs. They cannot recover or salvage the situation since they are implemented after the crisis. To say that they are reactive would be an overstatement—they are retroactive.

Failure to prevent business crises and inadequate resolution approaches sustain a large excuse-and-blame industry that creates absolutely no economic value—lawyers, claims experts, public relations consultants, arbitrators, expert witnesses, and assorted spin doctors. They depend on your failures for their livelihood and expect you to mismanage repeatedly. Human nature is preconditioned for failure; it is comforting, nonthreatening, and justifies expectations. In many companies today, a lot more effort is expended on responsibility avoidance, exculpation, and explanation than would have been required in real productive work to complete a project successfully in the first place. So what is to be done? The next time an emergency strikes, how do we find the gumption to pull ourselves together and avoid the urge to stick our heads into the sand, run for cover, or to sweep the debris under the rug?

The Proactive Crisis Resolution Approach

Do what you can, with what you have, where you are.

T. Roosevelt

As a first step on our journey to proactive crisis management, let us come to terms with the fact that most management problems are caused by people, directly or indirectly, and not by the circumstances or the "environment." People make things that break down, and cause negative "economic trends." Therefore, let us stop looking for external culprits, and "find the enemy," even if it is us. We will discuss the causes of crises in detail further in this text.

Although people are the primary instigators and causes of business emergencies and crises, positive actions such as hard work, perseverance, commitment, and decisiveness do not cause problems. It is the "bad habits" that do: inaction, sloth, incompetence, procrastination, indecision, and responsibility avoidance.

Breaking bad management habits is no different from breaking any other bad habit, addiction, or dependency. The willingness to do so is essential in order to break out of the vicious cycle of responsibility avoidance, failure, and denial. Like anyone who has tried to break a bad habit, a responsibility avoider and procrastinator should be aware of a few commonsense things to expect:

- You are in for a rude awakening.
- It is best to go cold turkey.
- You will have withdrawal symptoms.

Let's identify the basic steps for breaking our addiction to reactive crisis management and making a commitment to an active, resolution-oriented crisis-management approach.

1. **Recognizing/Coming to Grips with a Problem**

I believe that the American economy is in a fundamentally good shape. …

Herbert Hoover, 1930

The first hurdle in recognizing the problem is admitting that we in fact have a problem—getting out of the *denial* mode. Avoiders never get beyond this step, since the next step, taking responsibility, really hurts. In "speak-no-evil" companies, the corporate culture does not allow one to admit that there is a problem or to take it up the management ranks. The management expects problems to be resolved at lower ranks, thus turning the responsibility pyramid upside down—those at the top who should be the most responsible are the least responsible, sheltered by their underlings who absorb a disproportionate share of responsibility out of fear of repercussions. Nobody wants to be the messenger bearing bad news, and get shot.

2. **Taking Responsibility for the Problem**

The second step is taking *responsibility* for the problem—stop blaming others or the circumstances. As Napoleon said, there are no bad soldiers, only bad generals. If you are the general, then take responsibility—lead, follow, or get out of the way. Problems arise when people want to be generals but are incapable of leading, too conceited to follow, and too insecure to get out of the way.

3. **Devising a Solution**

Just admitting the problem's existence and taking responsibility is not going to solve it. Identifying the problem is half the solution, but only half. You must devise a solution to the problem.

4. **Implementing the Solution**

The solution is worthless unless it is implemented. Real problems get solved in practice, not in theory.

5. **Taking Responsibility for the Actions and Consequences**

Take responsibility for the problem and its solution; close the loop and let everyone know where the buck stops.

A great historical example of a clear-cut, simple, and action-oriented approach to resolving a crisis is Eisenhower's D-Day landing decision. He confronted a major problem head on: opening of a second front was essential to expedite a decisive victory in Europe. He listened to his advisers, took into account all the variables—the weather, enemy troop strengths and positions—kept his options open to the end, and then he made his decision and took personal responsibility for its consequences.

It is a little known fact that Eisenhower had a short handwritten note prepared, in case the invasion was a failure, to the effect that "the decision and responsibility for the failure of the invasion are mine alone." Note the essential ingredients in this action: make a *decision*, and take the *responsibility* for its consequences.

Eisenhower did not conduct an opinion poll, form a committee, or assemble a focus group to give credence to his decision or to serve as scapegoats in case of failure. In the end, each and every decision is made by an individual, and all responsibility is personal responsibility. Groups can study, advise, and recommend, but as Edison said, the genius lies in the lonely mind of a man.

Focus groups, committees, teams, and task forces of today's management fads are all too often just fronts for responsibility avoidance. We don't tackle the problems anymore; we raise issues.

By copying everyone on the memo that identifies a problem, we feel that it is no longer our problem once the memo hits the out-box. A committee becomes a giant swamp into which our personal responsibility is drained and diluted. Everybody's problem becomes nobody's problem.

Requirements for Proactive Crisis Resolution

How do we create an environment that is conducive to proactive problem resolution in our business organizations?

Following are what I see as the major requirements for a proactive crisis resolution approach:

- Competency.
- Urgency.
- Project mindset.

The Competency Requirement

I like studying the following three high-stakes yet common activities for their relevance to business emergencies: flying an airplane, running a hospital emergency room, and fighting a war.

Why? Two main reasons:

- High stakes.
- Low margin of error.

What do these three activities have in common? They all have very slim margins of error requiring very high competence to perform, system redundancies, and built-in problem absorption mechanisms. Redundancy in this context is a positive characteristic denoting the ability of a system to provide backups and reserves in case the primary functions fail. It is not a negative term implying useless, superfluous dead wood. Airplanes have multiple hydraulic systems, armies have reserves, and doctors have life-support systems at their disposal. The manner in which these activities are designed, organized, and managed—in other words their "systems dynamics"—will serve as an ongoing source of examples and references for project crisis management case studies, as well as for practical crisis management techniques, throughout this manual.

The earlier mentioned activities require competence and presence of mind just because the stakes are so high—the worst possible outcome is really, really bad. Precisely because these are such high-risk activities, we make every effort to ensure that they are accomplished successfully: we staff them with the best and brightest people and give them the best equipment and resources. Thus, good doctors save lives, good pilots don't crash their planes, and good generals win their battles.

If they aren't very competent at flying the plane/performing surgery/commanding troops, their passengers/patients/soldiers will die. And if they do survive to tell about their failures (at least the doctors, who don't normally die with their patients), they are quickly replaced by the more competent.

Why, on the other hand, is the business world rife and rampant with incompetence and failure when compared to this? I guess because we don't perceive the stakes as very high. If one company disappears, well, there are others that make more or less the same product or perform a similar service—which brings us to the second requirement.

The Urgency Requirement

In typical business situations, your life is not at stake unless, of course, you are in a high-risk occupation, such as mining, tunnel construction, oil drilling, or crime. In other words, there is no real urgency in becoming competent; being competent is not literally a matter of life and death. Urgency in the context of crisis management refers both to *time* and *importance*. It suffices to be just barely good enough to muddle through. Scoring As instead of Cs is perceived as a waste of effort.

What is the worst thing that can happen if we let the crisis run its course? We can lose our jobs and the company may go broke, but we'll most likely still be alive. There is a misconception that a business leader's incompetence does not affect the public.

It also helps that it is relatively easy to sweep business problems under the rug for a long, long time, especially in large, bureaucratic companies. The history of business is littered with examples of companies that failed "suddenly" and "unexpectedly," after recently being market and media darlings.

The lack of urgency mentality that allows management members to keep their heads in the sand and sweep problems under the rug can be summarized by the following two deadly characteristics:

- Perceived low stakes of the business risks.
- Environment conducive to crisis festering.

These two deadly sins must be expunged by instilling senses of urgency and importance to the prompt identification and resolution of crises. It is essential to approach the business crisis situations as if we have everything to lose: "the prospect of being hanged in forty-eight hours does wonderful things to focus the mind."

Project Mindset Requirement

In parallel with cultivating a habit of urgency when tackling crises, we must also get into the "project" frame of mind. A project is a mission—a specific objective to be achieved with limited resources, within specified performance parameters and a limited time frame. In other words, when a crisis hits, you won't have much time to resolve it.

The opposite of a project is a process—an ongoing repetitive operation with periodic performance goals, but no specific completion deadlines. Think of the examples of projects and processes.

Projects:

- Releasing new software.
- Developing a new product.
- Constructing a bridge.
- Flying a bombing raid.

Processes:

- Running a fast-food joint.
- Running a bank.
- Assembly line production.

If something goes wrong during a process, it can be stopped, analyzed, fixed, and put back into service with minimum disruptions to the overall operations or to the long-term objectives. A project, on the other hand, has to be completed without interruptions.

To draw an imperfect analogy, it is the difference between flying a plane and driving a car. If something goes wrong with your car while you are driving it, you can pull over to the side of the road, open the hood, and analyze the problem. The only loss will be a delay incurred to fix the car and in getting to your destination. Yet, if you encounter a crisis while flying a plane, there is no such luxury. You have to fly through the crisis—keep your plane in the air while trying to land without crashing. Crisis management, like all project activities, has an intrinsic continuity requirement: the overall "system" must continue to function during a crisis situation.

When you are faced with a crisis situation in your business, you will not have the luxury of suspending operations and sending the employees home while you tinker with the problems at your leisure. Business operations must proceed concurrently with crisis management, even if only the vital functions are maintained.

Although there is a significant school of thought in crisis management that insists on "shutting down" troubled projects until the problems are resolved, I strongly disagree. Doing nothing in most cases is a lot worse than doing something. In the meantime, someone has to pay for idle people and equipment, not to mention the significant remobilization cost premiums and learning curve costs that will be encountered during the restart. Furthermore, suspending the operation only gives us the opportunity to solve the problem theoretically—you have to resume operations to see if the fix works in practice. Again, if it can be stopped and resolved at leisure, it is a process-operation problem and not a crisis.

To summarize, all crisis situations are projects by nature and must be addressed with a project mindset. Whereas process improvement is continuous, incremental, and noncritical, crisis management is limited in time, scope, and objective. You have to start, sustain, and finish your crisis management task. To achieve that objective, you must identify the following:

- Objectives of crisis resolution: full recovery or preservation of the basic functions.
- Resources available for the resolution: time, money, people, etc.

PART I

Emergency Management

First Aid—Addressing the Symptoms

Emergency versus Crisis

In this book, both terms are used interchangeably; however, for the purposes of definition:

→ *Emergency* is a short-notice problem situation requiring immediate remedial action focusing on the visible symptoms.

→ *Crisis* is a protracted problem situation requiring action to cure the causes.

Again, to draw our medical analogy:

- *Emergencies* are *immediate* and require a first-aid/emergency-room treatment.
- *Crises* are *acute* and require intensive care treatment—most often a crisis is an emergency that has received a first-aid fix.
- *Problems* are *chronic* everyday glitches that may develop into full-blown emergencies and crises.

The overall approach to crisis management advocated in this book is clinical—that is, business problems are treated in the same intellectually rational manner that a doctor would treat an accident victim or a battlefield casualty. In order to apply the emergency and crisis management principles correctly, the Project Surgeon must understand the three phases of the process from crisis inception to cure:

- Emergency management.
- Crisis management.
- Crisis prevention.

Now that we have established the basic foundations for proactive crisis management, we can identify the dynamics of events that cause crisis situations.

Crisis Dynamics

In order to understand how a crisis develops and evolves, let us identify its major phases.

Phases of Emergency Management

1. Manifestation of symptoms.
2. Situation assessment.
3. Triage.
4. Emergency response.

Phases of Crisis Management

1. Manifestation of symptoms.
2. Identification of causes/definition of problem.
3. Collection of information required for developing a solution.
4. *Evaluation of alternatives and consequences.*
5. Solution development.
6. *Solution testing.*
7. Solution implementation.

Identifying the alternatives and consequences of a solution is part of a rational decision-making process, which will be addressed later in this text. In many situations, however, we have to rely on heuristics, or "rules of thumb" to expedite crisis resolution. The solution will not be perfect, but in a crisis, a good-enough solution will have to suffice, as long as it saves the life of a business. Search for a perfect solution is counterproductive if the business will bleed to death in the meantime.

This may be an appropriate time to revisit our analogy of emergency medicine, applied to a "sick" business. Think of crisis management as an illness or accident suddenly striking a company. As discussed earlier, curing the patient is a three-step process.

Immediate Action—First Aid/Emergency Room

Address the immediate symptoms only—stop the hemorrhage and seal off the damage. The patient is treated by the paramedics and brought into the emergency room. You will only perform a quick-fix, "emergency bypass."

Intermediate Action—Intensive Care

Once the patient is stabilized, address the causes behind the problems. You have now bought the time to perform crisis management Steps 1 through 7 in a systematic manner.

Long-Term Action—Corrective Surgery

As a corrective measure, the overall health of the business must be improved in order to eliminate the underlying sources of problems. Otherwise, the symptoms may temporarily retreat, but the business will remain highly susceptible and vulnerable to recurring crises.

Symptoms of a Crisis

As will be discussed in the Catastrophe Theory discussion in Part II, a crisis occurs when a congruence of events causes a business system to change from a "normal" to an "abnormal" state. Something goes haywire and pushes the normal, expected state of the world over the edge. Things become abnormal and don't behave as expected. There are two general ways in which the events cause a crisis situation.

Concurrent

Events that tip the scales from a normal to abnormal state act together, concurrently. This simultaneous occurrence of potentially catastrophic events may be manageable on a one-by-one basis, but together cause "system overload."

Successive

Events are successive and occur one after another, gradually weakening the business system with each blow. Each successive impact is amplified in a "domino effect."

How do we know when the normal state is about to change, and how can we identify the symptoms of a crisis in the making? There are four basic sequence modes of crisis symptom manifestation, as illustrated in Figure 1:

1. **Symptoms -> Crisis -> Failure**—Symptoms clearly indicate the cause.
2. **No Symptoms -> Crisis -> Failure**—In this case, symptoms may exist, but they are either compensated, ignored, or unidentified.
3. **Symptoms -> No Crisis -> Failure**—In this case, the failure is sudden and occurs very shortly after the first manifestation of symptoms, without a protracted crisis period. Symptoms are randomly manifested and do not indicate the cause of problems.
4. **No Symptoms -> No Crisis -> Failure**—This is a case of sudden, catastrophic failure without any visible symptoms or a crisis period. Problem festers with no obvious symptoms. The effect is delayed, and there are no early warning signs.

The first two scenarios are most common in business situations. Sudden, unexpected failures, as in mechanical systems, are rare. It is thus even more inexcusable if the symptoms are present but ignored, and the crisis festers untended.

Duration of each phase may vary. Often, the business system may compensate with "slack resources" and even mask the existence of a crisis. Since there are no warning signs, even a minor incremental increase in the severity of the problem can overload the system.

In addition to the gradual versus sudden modes of failure, we should also determine whether the entire system has failed or if just a component has failed and, if so, how critical the failed component is to the operation of the entire system. Some systems are unstable, with no redundancies—if the one critical thing fails, everything fails. A helicopter is a good example: if the rotor fails, the chopper falls.

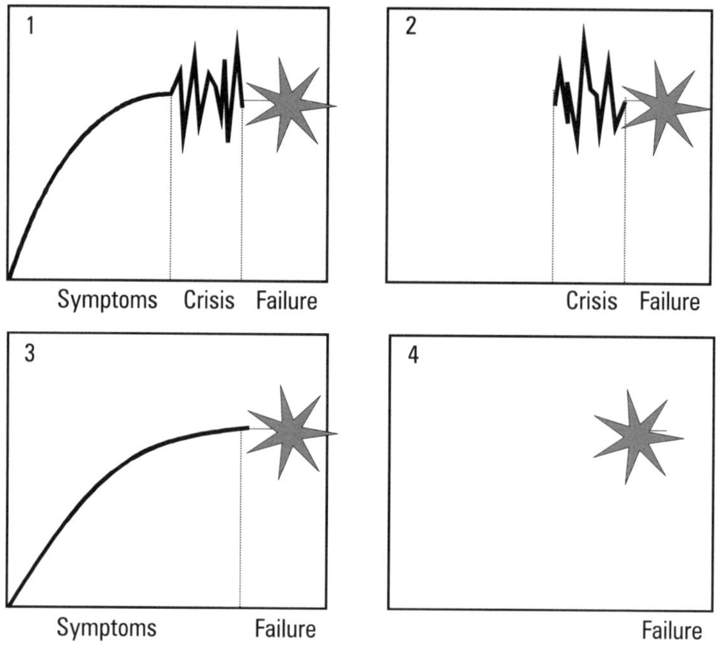

Figure 1. Crisis Manifestation Modes

In Part III, Crisis Prevention, we will talk about the management systems that must be in place to enable us to identify the symptoms of a crisis. Without knowing which symptoms to look for, and without a reporting system in place to alert us to the first appearance of symptoms, an onset of crisis will go undetected. It is important to note that crises symptoms are based on trends. A few random negative occurrences, or "outlyers," do not constitute a trend. It is a steady progression in a negative direction that should be a cause for concern. Let us look at some symptom examples in the project management context.

Individual project crisis symptoms:

- Decline in actual work-in-place compared to planned.
- Increase in indirect cost to direct cost ratio.
- Decrease in earned value/productivity.
- Increase in delays—actual time spent on a task versus planned.

Companywide project management crisis symptoms:

- Decrease in project revenue/net income per employee.
- Increase in budget overruns (actual costs exceed contract/bid amounts).
- Decrease in awarded projects as a percentage of total bid ("hit/miss ratio").
- Decrease in project backlog volume.
- Decrease in work-in-progress liquidity.

Needless to say, every project, company, or industry will have a different set of crisis symptoms.

Now we should be able to identify the symptoms of an emergency situation, understand how it develops, and have the basic skills to address it. The following case study will explain the thought process for identifying the crisis symptoms, assessing the big-picture status of a troubled project, and developing a recovery program. Then we will learn how to assess a crisis situation, in order to properly administer emergency measures.

A Case in Point—The Big Picture and Little Realities

The Project

"Here's the job," said the client, pointing at a desolate building site as the cab stopped, barely long enough to let the Project Surgeon out. "It's no big deal; fix it, and you'll be back in three months—I gotta catch the flight back now."

It was a nice spring day, and the site should have been swarming with activity. Instead there was no one in sight. The Project Surgeon made his way to the project office.

The atmosphere in the office was manic-depressive: eerie quiet punctuated by sudden spasms of activity caused by some real or perceived emergency. The staff members had long, despondent faces and the slouchy posture of a defeated army. The project was a medical building that should have topped-out by now. Instead, only the foundations were in progress. The project team had been on-site for almost a year into what should have been a two-year project, and just couldn't get any task started, let alone completed.

The Problem

It didn't take the Project Surgeon long to realize that the project had systemic problems on a large scale—this was not simply a case of delayed concrete deliveries or budget overruns. First of all, no one had either the responsibility or the authority to make a decision. Or, no one was exercising them. The job was underbid, but not to the extent that it couldn't have been bought out with some effort. However, the clients had set artificially low "target" buyout limits, lower than any low bids. Since no subcontractor was willing to drop the price to the "target," and the staff did not have the authority to award at offered bid value, trades were simply not awarded. By everyone's logic, they were just doing their job and being good corporate soldiers by not violating the rules. The clients stuck to their guns about award targets, and the staff did not award anything—thus, nothing much happened. Except that they all got their paychecks and bonuses, and otherwise managed to spend 25 percent of the indirect costs for about 5 percent of work in place.

It's not that they sat around doing nothing—problems were piling up daily. Every new phone call would send the staff members into a frenzy of activity trying to resolve the problem, only to be superseded five minutes later by another problem, and so on, until by the end of the day they were all exhausted having worked hard and accomplished nothing. The Project Surgeon realized that these people not only did not have the authority to do anything, but also they couldn't do anything even if they had the authority. They were chasing their own tails on a very slippery slope.

Many of the managers were capable and experienced people, but they were operating in a project environment without any system or structure. Their ability to manage and solve problems was severely hampered—on the one hand by the red tape imposed by the corporate bureaucracy, and on the other hand by a complete absence of supporting project management systems—and I mean a complete absence of any structure that is necessary to organize daily project management activities, right down to the basics. For example, there was no procedure for filing; thus, the paperwork did not get filed. Instead everything was piled, latest piece of paper on top, in what was called—and I am not making this up—a "to-be-filed pile." If the breakdown was on the basic administrative level, no wonder that the project management did not function.

The Resolution

The Project Surgeon got down to business immediately to try and ascertain the major problem areas. Specific problem solutions could not be developed without a larger, procedural framework in place. However, at the same time, he could not afford either to get bogged down in details or to embark on a complete procedural revamp. Thus, the repair approach was twofold: aggressively attack the ongoing problems, and concurrently develop a project management system.

First the Project Surgeon identified the major problem areas: a checklist comparing the actual status of things to what it should be under normal circumstances. With basic procedures in place, he had hoped that the problem discovery and resolution process would be institutionalized.

Table 1 shows the first action checklist/matrix. It seems straightforward in that it offers an action plan for all problems. However, developing the matrix was anything but easy, and its implementation was downright hard. First of all, only the most urgent and critical problems could make the list. They had to be burning issues that could materially influence the outcome of the job. However, they couldn't just be isolated random events—i.e., one-time catastrophes not emblematic of the big-picture problems. For the resolution plan to be effective, these problems had to be the manifestations of underlying causes. In regard to the action plan implementation, each item involved weeks of tedious work to rebuild the project management infrastructure. For example, with regard to the five action items under Subcontract Awards, the Project Surgeon had to start from scratch, by first locating the bids for all trades, recreating the budget, and so on.

In addition to the detailed recovery action plan, the Project Surgeon condensed its main points into an executive summary, so that the clients would be aware of the magnitude of the effort required to recover the project.

Project Recovery Plan

1. Complete existing condition survey, and determine surveying and construction-tolerance coordination requirements for all building systems and components.
2. Prepare a complete project schedule including buyout, procurement, and construction activities. Evaluate time spent versus work completed, and reschedule accordingly.
3. Set the accounting system, including budget, target-buyout values, general conditions, cash flow, billing cycle, anticipated cost report, etc. Evaluate money spent versus work completed, and rebudget accordingly.
4. Set a computerized document management system to process all submittals, transmittals, requests for information (RFIs), changes, etc.
5. Review all drawings and specifications, and submit change order requests for all missing, incomplete, or incorrect items.
6. Review and update all documentation submitted to date, including requests for proposal and RFIs, and submit claims for all valid extra work.

7. Develop a standard subcontract document package to ensure consistency and flow down of general contract terms.
8. Develop a detailed submittal and long-lead item-procurement schedule per the client's requirements. Set required submittal turnaround and material/equipment delivery dates in subcontracts.
9. Review submittal and procurement items processed to date, and push delayed items.
10. Rebid, review scope, and buy out critical trades.
11. Develop a project procedures manual to ensure compliance with client's standard formats and procedures for submittals, changes, and other paperwork.
12. Set a project quality-control program in accordance with the client's requirements.
13. Set required daily/weekly/monthly reporting formats.
14. Set field coordination procedures for major trade groups.
15. Monitor progress on a regular basis, and take timely corrective action.

Conclusion

The project turned out to be a relative success. I would term it a *salvage operation* rather than a *recovery*, since the original budget and schedule goals were not achieved. One can only speculate that the results would have been far worse if no recovery plan had been attempted. The main fault was that the plan started too late—after the point of no return had passed—to fully influence the final outcome.

Situation Assessment

One of the major obstacles in resolving a crisis situation is defining exactly what constitutes a crisis. We know things are not going well, but compared to what benchmark? Without a clearly defined yardstick, we will not be able to objectively evaluate our performance. If our performance is "bad," then what exactly constitutes a "good" performance? If we are "late," then what does "on schedule" mean?

In order to answer these questions objectively, we must first define our objective. We must know what it is that we are setting out to accomplish in the first place before we can judge how well or how poorly we are performing in comparison to the established objective. We must also have a plan—a road map defining how the objective will be accomplished, with interim "markers" or milestones to warn us if we are straying off course. An objective and a plan establish a clear performance target, as well as a basis for quantifiable performance comparison.

Comparison of the current actual status and the current planned status gives us a snapshot of our performance—what should be happening versus what is happening. How difficult it is to bridge the gap defines the severity of the crisis. A crisis is a set of circumstances that throws us off course to our objective. A crisis situation must be identified by warning signs or red flags, which identify a significant gap between the planned and actual performances.

Problem	Symptom/Status	Action
Subcontract Awards		
Subcontract Award Status	Only three trades awarded; no follow-up.	Award all trades ASAP at the best financial terms possible.
Trade buyout/subcontract award spreadsheet showing overruns/underruns.	Nonexistent: financial impact of awards not known until entered in accounting system.	Develop a budget-versus-award comparison matrix; check status weekly.
Target award dates/buyout schedule.	Not updated.	Enter drop-dead dates for awards; check status weekly.
Trade coordination included in subcontracts; e.g., steel/curtain wall.	Incomplete.	Write into subcontract language, and negotiate inclusion in scope.
Check for missing coordination items in awarded subcontracts.	Incomplete.	Same as above.
Submittals		
Submittal schedule deadline included in subcontracts.	Partial.	Same as above.
Projected submittal schedule for each trade showing quantity and dates.	Nonexistent.	Develop listing of all submittals, by vendor, with submittal dates; monitor and expedite weekly.
Submittal tracking system on computer database.	Nonexistent; submittals that arrive are sent randomly for review, or back to vendors.	Set up Prolog software; enter processed submittals retroactively.
Turnaround delay reporting/expediting procedure in place.	Nonexistent; submittals are processed if and when they arrive.	Same as above; check current status, and expedite all late items.
Submittal review/coordination procedure in place.	Nonexistent; each project manager randomly processes submittals.	Develop standard review procedure, and ensure compliance by all staff.
Procurement		
Procurement schedule deadline included in subcontracts.	Partial.	Write into subcontract language, and negotiate inclusion in scope.
Procurement tracking/expediting procedure in place.	Nonexistent; nothing is being fabricated or manufactured yet.	Issue written procedure for expediting procurement from submittal phase through fabrication phase.
Material and equipment expediting database in place.	Nonexistent.	Coordinate long lead fabrication with overall schedule, and develop tracking procedure based on on-site required dates.
Vendor/supplier/fabricator directory on file.	Nonexistent.	Develop an organization chart/hierarchical bill of material for each component.
Shipping/warehousing/delivery procedures in place.	Nonexistent.	Develop shipping, delivery, and storage procedures. Perform inventory requirement analysis simulation.
Fabrication facilities visited and fabrication verified.	Not done.	Arrange for regular expediting/verification trips.
Subcontract Management		
Executed subcontract binder current and on file.	Not updated.	Compile all paperwork in one place.
Change order request log current and on file.	Nonexistent; change orders not pursued with client.	Audit all scope revisions, and retroactively price and issue change orders.
Change order expediting procedure in place.	Nonexistent.	Use standard software to track status and commitments.
Subcontract and owner changes cross-referenced and current.	Nonexistent.	Same as above.
Project status log updated, showing execution status.	Not updated.	Compile all status information in one summary matrix, and update weekly.
Cost Control		
Anticipated cost/projected cost-to-complete procedure in place.	Incomplete: accrued status/spent money reported only.	Develop cost-to-complete projections based on current commitments; update monthly.
Profit/loss and earned value shown in cash-flow reports.	Nonexistent; six monthly billings issued to date.	Reconcile spending versus money received; update monthly.
General conditions/indirect cost cash-flow profit/loss report.	Nonexistent.	Reconcile spending versus money received; update monthly.
Billing and payment cycle Accounts Payable (AP)/Accounts Receivable (AR) report.	Nonexistent.	Compile all accounting data, and develop report.
Schedule and Logistics Plan		
Weekly and monthly schedule status update.	Occasional updates/inaccurate.	Ensure correct status reporting and expediting of late items on a weekly and monthly basis.
Site logistics plan: staging, layout, access cranes, hoisting, etc.	Nonexistent: equipment located randomly.	Develop physical plan, and organize site accordingly.
Construction Operations		
Existing condition survey.	Not done; don't know if project is being built to specified dimensions and tolerances.	Resurvey completed work to determine if new work will "fit"; repair existing tolerances when feasible.
Structural coordination with curtain wall.	Poor.	Coordinate and rewrite subcontract scope language.
Structural coordination with Mechanical/Electrical/Plumbing (MEP) Systems.	Poor.	Coordinate and rewrite subcontract scope language.
Construction sequencing/cycle analysis/linear balance program.	Not done; work performed randomly.	Develop productivity parameters and activity cycles; balance production in the field.

Table 1. First Action Checklist/Matrix

1. Define the original objective. What is it that we are supposed to achieve, and by when?
2. Define the plan to achieve the objective. What are the specific means and methods that will be implemented to reach our goal?
3. Break down our plan components into a set of smaller, interim objectives with interim milestones that cumulatively lead to our goal.
4. Check the current status of events at each milestone. Where should we be now if things were progressing as planned? How bad is the current gap?

Periodic performance assessment at regular intervals allows us to identify and correct problems before they magnify into a full-blown crisis. All projects, missions, and expeditions are planned like this. Businesses file quarterly financial reports and monthly receivable aging reports. Any problems identified at these interim checkpoints will be smaller and more manageable than if they were evaluated at longer intervals—say, annually. Mountain-climbing expeditions have a number of base camps on the way to the summit and back. If contact is lost, then the rescue mission knows that it has to focus on the area between the last contact checkpoint and the next one. Generally speaking, if the performance is progressing according to plan, fine; if not, we are at a crossroads. A number of questions must be answered, among others:

- ✔ Can we get back on course, and still make our objective?
- ✔ What will happen if we do nothing?
- ✔ How do we get back on course?

To answer these questions, we must determine the magnitude of the problem and conduct a *realistic assessment of recoverability*. This means that we have to honestly assess whether we can still get back on track and conceivably achieve our original objective. Or, are we past the "point of no return" where all we can hope for is saving the situation the best that we can?

Look at Figure 2, which represents performance of a project over time with the ultimate goal of reaching a Target objective at point 1. Performance must be defined: profits earned, widgets produced, percent of project completed—i.e., the main objective of the project. The breakeven point (2) represents the last chance to fully reverse the crisis and recover the project to its original target.

Sometimes it is possible to reach the original target after passing the breakeven point, but only with additional effort and resources—usually by buying time with money. Note in Figure 2 how the slope of the recovery line gets steeper the further away we move to the right of the breakeven point. The longer we wait to implement recovery action, the harder it is to get back on track. After a certain point, it becomes impossible to jump back to the recovery line—the effort slope is too steep. We will call this point "the point of no return." If no additional resources are available, then the breakeven point becomes the point of no return, as is often the case.

If we wait until we reach the point of no return (4 in Figure 2), the best that we can hope for is to achieve point 5—significantly below our original target. If we do nothing, the situation could deteriorate to point 3, which could be disastrous.

The Added Effort line in Figure 2 represents the maximum additional effort that can be expended—i.e., the steepest slope. Thus, if you draw a parallel Added Effort line after the point of no return, it will not connect to point 1, the Target objective.

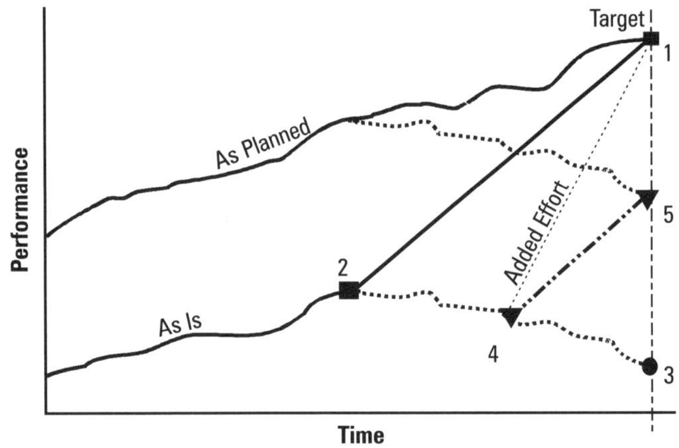

Figure 2. Assessment of Recoverability

In a crisis situation, the point of no return has usually been passed, and the project cannot be brought back to its originally intended target without added effort. Some aspects not directly related to the project objective must be sacrificed, and additional resources must be expended.

If we are past the point of no return, to what degree can the project be salvaged or recovered? Establish clear and realistic achievable goals; the longer you wait, the less that can be done.

The Assessment of Recoverability and Recovery Plan

Define Your Objective

Determine if you are dealing with a *recovery* or a *salvage* operation. Ask the following questions:

- ✔ What is the worst possible outcome? What is at stake?
- ✔ What will happen if nothing is done?
- ✔ Can the project be brought back to the originally planned target, or has the point of no return been passed?

Depending on your answers, you may define the situation as follows:

- ➔ If a crisis plan is implemented before the breakeven has been reached, then it is a recovery operation.
- ➔ Between the breakeven point and the point of no return, recovery may be feasible by expending additional effort.
- ➔ If a crisis plan is implemented after the point of no return has been reached, then it is a salvage operation.

Realistically identify your objective: is it a case of simply cutting your losses—i.e., project completion at minimum financial loss in the shortest possible time—or is it a matter of saving your company or project's life.

Focus on What Counts

You won't have time to solve, let alone tackle, all the problem issues. In order to be effective, you must allocate your resources and prioritize your goals. Not everything is equally important. You must learn to cut your losses and pick your battles. More will be said on this topic in the next two sections, but for now, try to envision a short list of essential items that absolutely must be addressed in order to save the situation. Everything else can burn down.

- ✔ Identify decisions that must be taken and executed—*sine qua non*—lowest common denominator.
- ✔ Prioritize problems by the relative value of their solutions.
- ✔ Rank problems by magnitude/outcome.
- ✔ Rank problems by urgency.
- ✔ Allocate resources to solvable, important problems only.
- ✔ Allocate limited resources to problems that are most likely to be solved and most worth solving.

Lowest-Total-Cost Concept

A brief discussion of this economic concept is important in order to drive home the point that all solutions, and emergency solutions in particular, are compromises between conflicting objectives that must all be satisfied to a certain extent. Seldom will there be a perfect solution to an emergency problem. Rather, you will be faced with a series of tradeoffs: in order to get some of this, you have to give up some of that. The tradeoff process is discussed further in the decision-making discussion in Part II. Here we will try to focus on achieving the best overall solution to an emergency, basically by asking ourselves the following questions:

- ✔ What is the cost to fix the problem versus the cost not to fix it?
- ✔ What is the solution worth to me?

The dynamics of any system, from an automobile to a corporation, requires constant balancing between the conflicting demands of the cost of failure versus the cost of perfection. The objective is to minimize overall/total cost of a system—minimize the cost of perfection and the chance of failure.

For example, your car has thousands of parts with a wide range of useful life and probability of failure. All will fail after a certain period of time. You want to keep your car as long as possible and wish to avoid costly repairs. You can spend three hours a day maintaining your car engine and be pretty sure that no major part will fail due to your vigilance. But within say, ten days, the cost of your maintenance (in time, oil, parts, etc.) will exceed the replacement cost of a part whose "time has come" to fail. The cost of preventive maintenance has exceeded the cost of replacement. Thus, the "preventive maintenance schedule" says to change the oil every 3,000 miles or every three months, not every day. Less frequently than that significantly increases the probability of engine breakdown; more often does not yield additional benefits.

There are three basic methods of maintenance:

- ➔ **Preventive**—Maintenance is time based and performed on a regular time schedule.
- ➔ **Reliability-Centered Maintenance**—Focus on the performance of each component, and maintain exactly when needed. Frequency could be more or less often than prescribed by a preventive schedule. Objective is to maintain critical parts first—those parts that affect the reliability of the entire system.

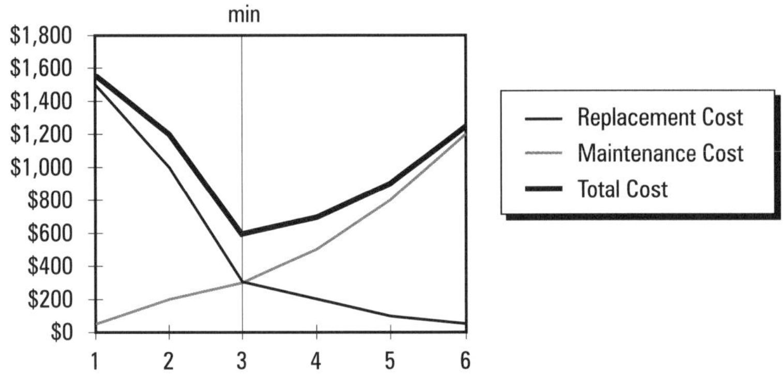

Time	1	2	3 (min)	4	5	6
Replacement Cost	$1,500	$1,000	**$300**	$200	$100	$50
Maintenance Cost	$50	$200	**$300**	$500	$800	$1,200
Total Cost	$1,550	$1,200	**$600**	$700	$900	$1,250

Figure 3. Lowest Total Cost

➔ **Run to Failure**—Some things are worth maintaining, others are sometimes cheaper to replace (shoes, VCRs, toasters), or are not designed to be repaired (light bulbs).

Why is the lowest-total-cost concept important in business crisis management?

- Prioritized resource allocation: time, money, manpower.
- Survival of the fittest.
- Excellence, not perfection.
- Maximum results with minimum efforts.
- Understanding why "things break down"—a diagnostic tool.

The application of the lowest-total-cost concept is a basis of all decision tradeoffs, negotiations, bargains, and compromises. To get something, we usually have to give something. What are the acceptable losses and unacceptable costs? The trick is to find the right balance. Here are some examples:

- Cost of safety versus cost of accidents.
- Cost of quality versus cost of rework.
- Cost of maintenance versus cost of replacement.
- Cost of time versus cost of money.

Understanding this concept allows us to evaluate the cost versus the value of fixing a problem, and ensures an effective application of triage.

A Case in Point—Project Recovery

We're losing money on most of our projects, but we'll make it up on the volume.

A project manager of two similar building projects (about $90 million budget, twenty-four-month schedule each) made this true statement. The first project was nearing completion at two months behind schedule and 35 percent over budget. The budget overruns and delays were not discovered until it was too late to take any corrective measures. Financial problems were not noticed until the eighteenth month when the budget was exceeded. The awareness of delays dawned on the project team when twenty-two months had elapsed and the project was still not done. The second project was just starting and, outwardly at least, had not yet developed any budget or schedule problems.

Since the first project was almost complete, there was very little that could be done to recover, but at least we could analyze what was going on and develop recovery procedures for the second project. The Project Surgeon noticed that the major problem was that there was no indication that anything was going wrong during the course of the first project—individuals only knew that they were in trouble after they got in trouble; twenty-four months had passed and $90 million was spent, and they still weren't done.

The usual day-to-day problems notwithstanding, nobody on this project had asked a simple question: Is my project on schedule and on budget *now*? This question should be asked and answered on a regular basis.

Before getting into the specifics of the first project, the Project Surgeon sat down with the management team to ensure that we had a consensus regarding some basic project budget concepts. The way this case is presented seems simplistic to a seasoned project manager; however, getting to the root cause of the problem—reducing the problem to its "lowest common denominator"—was essential before we could attempt an organized recovery effort. This narrative walks the reader through the process that the Project Surgeon undertook to identify problems and establish recovery parameters for the second project that this team was managing. Various team members' comments are paraphrased (bolded here) to guide the discussion.

All project budgets consist of three components: direct cost, indirect cost, and profit. There are different names, variations, and subcomponents, but for all practical purposes, any project budget can be reduced to these three components.

Direct cost (also known as trade costs) is work directly performed to achieve the project—e.g., pouring concrete, attaching plane wings to fuselage, writing computer code—whether performed by your own workforce or subcontracted. All direct cost budgets are based on productivity measures and unit costs.

Indirect cost (also known as general conditions or overhead) is the management cost required to complete the direct work. Indirect cost should be an-as-small-as-possible percentage of direct costs—i.e., the ratio of watchers to workers and paperclips to steel beams should be as small as possible, but not too small so that the project cannot be controlled.

One of the main causes of budget overruns and schedule delays is inefficient management or "high overhead." Management is a fixed cost—managers "manage" even if no real work is performed; they are not sent home if it rains.

If the indirect cost budget is 10 percent of the direct costs, then this ratio has to be 10 percent or better for each fiscal period of the project.

Profit (or loss) can be a separate category, or it can be expressed as any money we save (or lose) on the direct and indirect costs.

I understand all that, but the problem is that we just didn't know that we were over budget until it was too late to do anything about it. How can we monitor our financial situation on the project on a regular basis?

Once the basics of the project budget components were understood, we were ready to introduce the concept of cash flow. The pervasive backdrop of any crisis is the lack of money. In a sense, every crisis is a financial crisis. To a varying degree, lack of money causes or exacerbates a bad situation; influx of money cures it.

Managing money during a crisis is probably the most important crisis management skill. It's not the overall lack of cash that kills a project; it's the lack of cash when we most need it—in other words, lack of cash *now*.

Managing money for a project or a corporation is relatively simple, assuming that you can manage your own money well. Not to sound overly simplistic, when we eliminate the background noise of highbrow financial terminology, tax lingo, and other distractions, managing money well boils down to a simple formula: cash in less cash out equals profit. You sell something to someone (goods, services); this is your revenue, not to be confused with income. You buy something—i.e., pay someone—whether or not you like to do so (supplies, rent, raw materials, payroll, taxes). These are your costs. The difference between the revenue and the cost is your net income or profit. The only way to increase your profit is to increase your revenues and/or decrease your cost.

If you can balance your own checkbook, you should be able to balance your company checkbook. Things get complicated when the factor of time is introduced. Unfortunately, breakeven analysis does not account for the discrepancy in time between selling a product and getting paid for it, and the cash required to sustain you between these two points in time.

It is the cash flow, not the cash. In other words, it is the liquidity or the ability to meet current obligations—having the cash influx to pay your bills when they are due—not overall solvency, that is paramount. You can still be solvent but not liquid. Your assets may cover your liabilities, but you may be short of cash to pay your bills.

Yes, but other than opening a separate checking account for each project and not paying your bills when the checks will bounce, what can a project manager do to anticipate and control his project's cash flow?

Establish a financial cycle. Ensure positive cash flow when the bills are due. The fiscal cycle of all project accounting procedures is based on the progress payments for work completed. There are three main reasons why a monthly fiscal cycle is the norm for proper project accounting:

- The progress payments for value of work completed on any project are governed by the contractor's financial obligations to its vendors, suppliers, employees, and lending institutions. To maintain adequate cash flow of operating funds, the contractors almost always submit invoices for progress payments on a monthly basis.
- Regular and timely accounting data input is required for proper cost performance control by the project management. If the accounting data is accrued and processed in less regular intervals—say, on a quarterly basis—it may be too late to take adequate cost-control measures.

- Project cost accounting differs from manufacturing, "work-in-process" type accounting, and service-industry, "billable-hour" type accounting in that there is an ongoing need to anticipate future costs on projects on a regular basis. Whereas in the manufacturing industry, the actual costs are accrued as the product is built, overhead and profit is added, and product priced accordingly, a project manager in essence promises to complete a project at a future date for a previously promised sum of money. Since the project costs are so much harder to control, forecasting on a regular basis and based on current accounting data is essential to take corrective measures as soon as problems occur.

It seems that this process requires balancing payables and receivables on a monthly basis, and then projecting the cost to complete the project.

That's right, but you must avoid the mistake of mixing cash and accrual accounting—cash for payables and accrual for receivables. You cannot balance real obligations with fictitious income. We tend to overstate future income and receivables in order to balance hard, current, actual liabilities. We invent the future: windfalls, tax refunds, claim settlements, and sales increases in order to make up for the money already spent. Look at any corporate business plan or *pro forma* income statement—you never see a decrease in the projected sales and income, only a steady, optimistic growth upward.

The same principle applies to the long-term assets and liabilities. Ever notice how assets are sold at 10c on the dollar, while the liabilities just won't decrease. Be realistic.

The best way to ensure positive cash flow is to make sure that your money is working for you. Are you going to run out of money before you finish your project and, worse yet, before the big payout is due?

So what you are saying is that because the budget is essentially a target on the horizon, we can't use the traditional look-backwards accounting to manage projects. We are all familiar with the time-value-of-money concept, but I don't see how it would help in this case.

If the *cash* in the words *cash flow* is about money, the *flow* part is about time. Most engineering and project management professionals are familiar with the concept of "time value of money." They understand net present value, yields, and other investment and return measures. While these techniques are certainly useful from the standpoint of project financing and life-cycle cost analysis, they don't help us monitor and control the projects in real-time mode. Unfortunately, very little attention is paid to the concept of "earned value" or integrated cost/schedule control from a project standpoint.

What do you mean by that?

Well, all things being equal, there are only four possible cost and schedule outcomes on any project (along with construction superintendents' colloquial words for each):

1. Under Budget and Under Schedule—"Hero Job."
2. Under Budget but Over Schedule—"Cheap Job" or "Slow Job."
3. Over Budget but Under Schedule—" Fast Job" or " Expensive Job."
4. Over Budget and Over Schedule—"Bum Job."

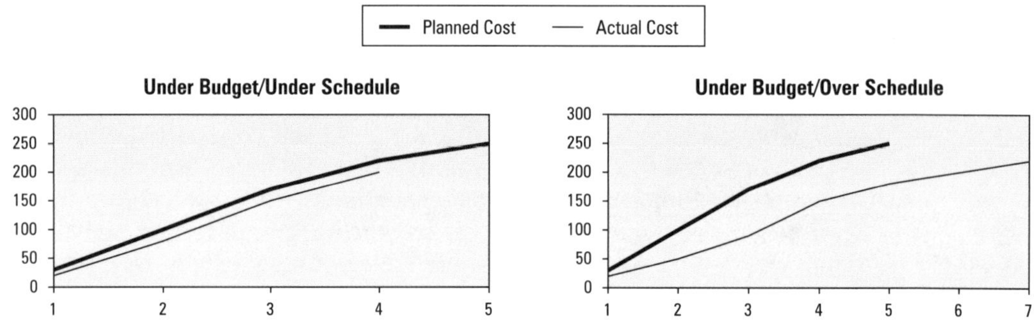

Period	1	2	3	4	5	6	7
Planned Cost	30	100	170	220	250		
Actual Cost	20	80	150	200			

Period	1	2	3	4	5	6	7
Planned Cost	30	100	170	220	250		
Actual Cost	20	50	90	150	180	200	220

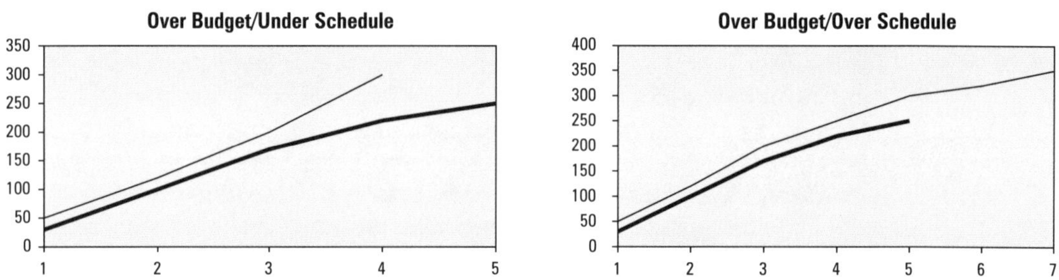

Period	1	2	3	4	5	6	7
Planned Cost	30	100	170	220	250		
Actual Cost	50	120	200	300			

Period	1	2	3	4	5	6	7
Planned Cost	30	100	170	220	250		
Actual Cost	50	120	200	250	300	320	350

Figure 4. The Four Possible Project Outcomes

If we plot budget/cost on the y axis and time on the x axis, these outcomes can be graphically represented as shown in Figure 4.

Your first project is a bum job. Your second project is only five months into the schedule, and it could turn out to be any of the four scenarios. Of course, the intent is to bring in a hero job. If your current project status is any of the other three, then bringing it back to on schedule and on budget will be a full recovery; anything less will be salvage. For example, if the project is currently both over budget and behind schedule, the recovery sequence can be 4 ->3 ->1 or 4 ->2 ->1.

Great, but before we can recover or salvage the project, we need to know that we are behind schedule and over budget. How can we know that on an integrated basis?

That's right. We must first establish a planned cash-flow distribution—or a planned expenditure of project costs, before we can start monitoring the actual expenses.

An integrated cost/schedule control system establishes planned budgets and schedules, and also allows for monitoring actual cost and schedule performance. This is done by measuring expenditures against budget and against earned value, and by identifying variances so that corrective action may be taken when required.

OK, how do we develop an initial cash flow that will serve as a benchmark for measuring actual performance?

First, develop a programmatic budget and schedule for the project. Ideally, each budget-line item should be represented by a corresponding schedule item.

Spread budget costs over the estimated duration for each activity. The spread will most likely not be in equal installments, since cost loading over time is dependent on numerous variables. Usually it assumes a bell-shaped normal distribution curve with activity costs building up toward the middle of the duration, then tapering off. Note that the total project cash-flow curve is for the sum of all activities and is usually also bell-shaped. Cumulative cost curve will be S-shaped.

The earlier-mentioned steps can be completed using off-the-shelf scheduling and spreadsheet software. Some software packages, such as Primavera, have an integrated cost/schedule generation capability with cash flows generated and modified automatically, based on the changes in cost, schedule, or cost-loading input.

It is important to note that the baseline cash flow—i.e., the budget and the schedule—should not be adjusted during the course of the project unless the project scope or the schedule are revised, since it must serve as a benchmark for cost and schedule performance.

Now, let us establish the cost measures, variances, and indexes that are used to track cost and schedule performance on an integrated basis in "real time." The basic method has been around for a couple of decades and is commonly known as the *cost/schedule control system criteria (C/SCSC)*. Here are the main characteristics of this method:

- It sets time/cost progress monitoring and performance standards.
- It allows for measuring expenditures against budgets.
- It allows for identification of variances in order to take corrective action (scope changes, contingency adjustment, budget/schedule adjustments).

The key cost measures can be summarized as follows:

- Budgeted cost of work scheduled (BCWS) is the budgeted amount of cost for the work scheduled to be completed in a given time period. It answers the question: "How much money should have been spent if work had been completed on schedule?" This is the baseline budget.
- Actual cost of work performed (ACWP) is the amount actually expended in completing an item of work during a given time period. It can be more, less, or equal to the budgeted cost.
- Budgeted cost of work performed (BCWP) is the budgeted amount of cost for the work completed in a given time period, also known as the Earned Value of work accomplished. BCWP is equal to BCWS multiplied by the actual percent of work completed. It answers the question: "How much money should have been spent for work actually completed?"
- Cost Variance = BCWP – ACWP (+ is under budget)
- Schedule/Performance Variance = BCWP – BCWS (+ is on schedule)
- CPI (Cost Performance Index) = BCWP/ACWP (>1 is under budget)
- SPI (Schedule Performance Index) = BCWP/BCWS (>1 is on schedule)
- Cost Percent Overrun (underrun) = ((ACWP) – (BCWP))/(BCWP)

The use of both cost and schedule variances and indexes calculated in this fashion provides an integrated cost/time-reporting system that measures cost performance in relation to the work accomplished, and ensures that both time scheduling and cost budgeting are constructed upon the same database.

Now let us look at the worst possible project outcome, briefly described earlier, in complete detail, with all indexes and variances shown on a monthly basis (see Figure 5). Note that your first project is the basis for Example 1—over budget and behind schedule. Your second project is only four months old. Let us assume that the cost and schedule are the same as for the first project. Note also that the monthly actual percent complete and the ACWP line item values for month 5 and beyond are projected values used to illustrate the recovery process.

There are two basic methods of recovery. The first method relies on the C/SCSC. It involves taking proactive measures necessary to improve the cost and schedule performance; it is a performance and productivity approach.

The second is a financial approach. It involves techniques required to maintain constant project solvency—i.e., positive cash balance for each fiscal period.

Project Recovery—Performance/Productivity Approach

Example 2 shows us how to recover from a bum job to a hero job using the productivity approach. First of all, note that I have added several performance recovery measures not commonly used in the project management literature, along with their application in the recovery process (see Figure 6, pages 37–40).

Budget variance = BCWS – ACWP. Negative cost variance (earned value) assumes continued poor performance. If the recovery measures are implemented, the earned value can be brought to budget—i.e., BCWP = BCWS—then it is sufficient to recover to the original budget.

Completion variance = actual % complete – planned % complete. The cumulative completion variance in this example is positive through the third month, which means that we are accomplishing more work than originally planned. But then it slips below zero and gets progressively worse.

Schedule recovery requirement = completion variance x total scheduled periods. Schedule recovery requirement (negative months) is calculated as completion variance times the total number of months and answers the following question: "How much time do we have to recover *now* in order to recover the overall schedule?"

Schedule recovery index (%) = schedule recovery requirement/periods remaining. Schedule recovery index (%) is calculated as schedule recovery requirement divided by number of months remaining. It is the true measure of required recovery effort and its feasibility. For example, in the twenty-first month, we have to recover 1.81 months in the remaining three months of the project—i.e., we have to complete 4.81 months' worth of work in three months, or add 60.44 percent more effort/manhours/resources. It does not seem feasible. In the twenty-third month, the index is 117.33 percent, indicating that we have to recover over one month in the last month remaining, which is impossible. Thus, sometime between the twenty-second and twenty-third month, we have reached the schedule recovery point of no return, and delays become inevitable, assuming we have unlimited resources. In reality, we will run out of resources (time/money/manpower/materials) long before.

Base Cost and Schedule Parameters

Total Project Cost (× 1,000)	$90,000
Project Duration (months)	24
Average Monthly Cost (× 1,000)	$3,750
Maximum Monthly Cost (× 1,000)	$5,000

Cost Measures:

Budgeted Cost (BCWS) Budgeted Cost of Work Scheduled—Shows Base Budget Spread Over Base Schedule ($90 Million in 24 Months)
Earned Value (BCWP) Budgeted Cost of Work Performed—Shows Base Budget Spread Over Actual Schedule ($90 Million in 26 Months) = BCWS × Percent Complete
Actual Cost (ACWP) Actual Cost of Work Performed—Shows Actual Costs Expended Over Actual Schedule ($121.5 Million in 26 Months)

Monthly Cost Cash-Flow Analysis

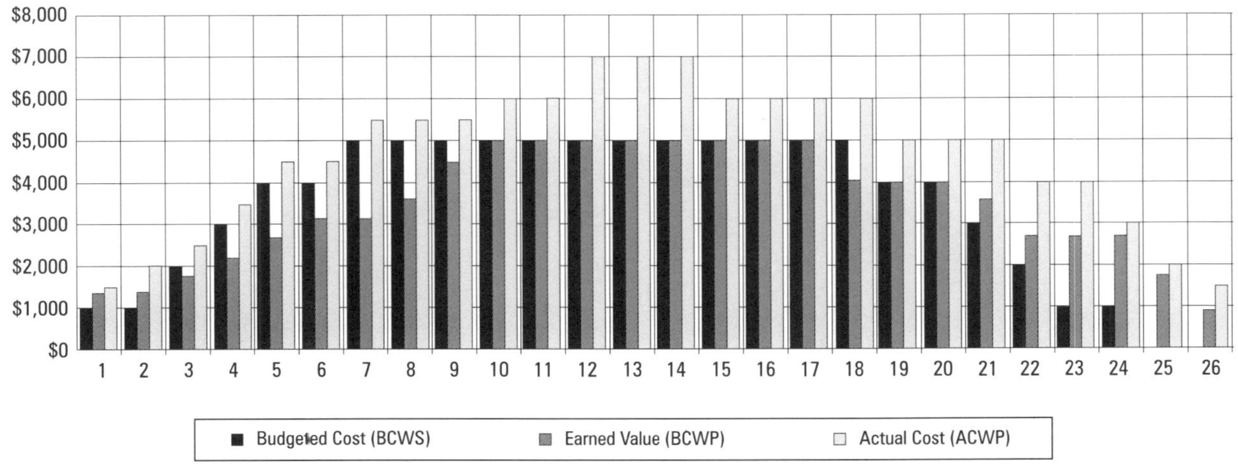

	1	2	3	4	5	6	7	8	9	10	11	12	13
Budgeted Cost (BCWS)	$1,000	$1,000	$2,000	$3,000	$4,000	$4,000	$5,000	$5,000	$5,000	$5,000	$5,000	$5,000	$5,000
Monthly Percent Complete	1.50%	1.50%	2.00%	2.50%	3.00%	3.50%	3.50%	4.00%	5.00%	5.50%	5.50%	5.50%	5.50%
Earned Value (BCWP)	$1,350	$1,350	$1,800	$2,250	$2,700	$3,150	$3,150	$3,600	$4,500	$4,950	$4,950	$4,950	$4,950
Actual Cost (ACWP)	$1,500	$2,000	$2,500	$3,500	$4,500	$4,500	$5,500	$5,500	$5,500	$6,000	$6,000	$7,000	$7,000

	14	15	16	17	18	19	20	21	22	23	24	25	26
Budgeted Cost (BCWS)	$5,000	$5,000	$5,000	$5,000	$5,000	$4,000	$4,000	$3,000	$2,000	$1,000	$1,000	$0	$0
Monthly Percent Complete	5.50%	5.50%	5.50%	5.50%	4.50%	4.50%	4.50%	4.00%	3.00%	3.00%	3.00%	2.00%	1.00%
Earned Value (BCWP)	$4,950	$4,950	$4,950	$4,950	$4,050	$4,050	$4,050	$3,600	$2,700	$2,700	$2,700	$1,800	$900
Actual Cost (ACWP)	$7,000	$6,000	$6,000	$6,000	$6,000	$5,000	$5,000	$5,000	$4,000	$4,000	$3,000	$2,000	$1,500

Figure 5. "Bum Job"—Project Completed Over Budget and Behind Schedule

Continued on next page

Cumulative Cost Cash-Flow Analysis

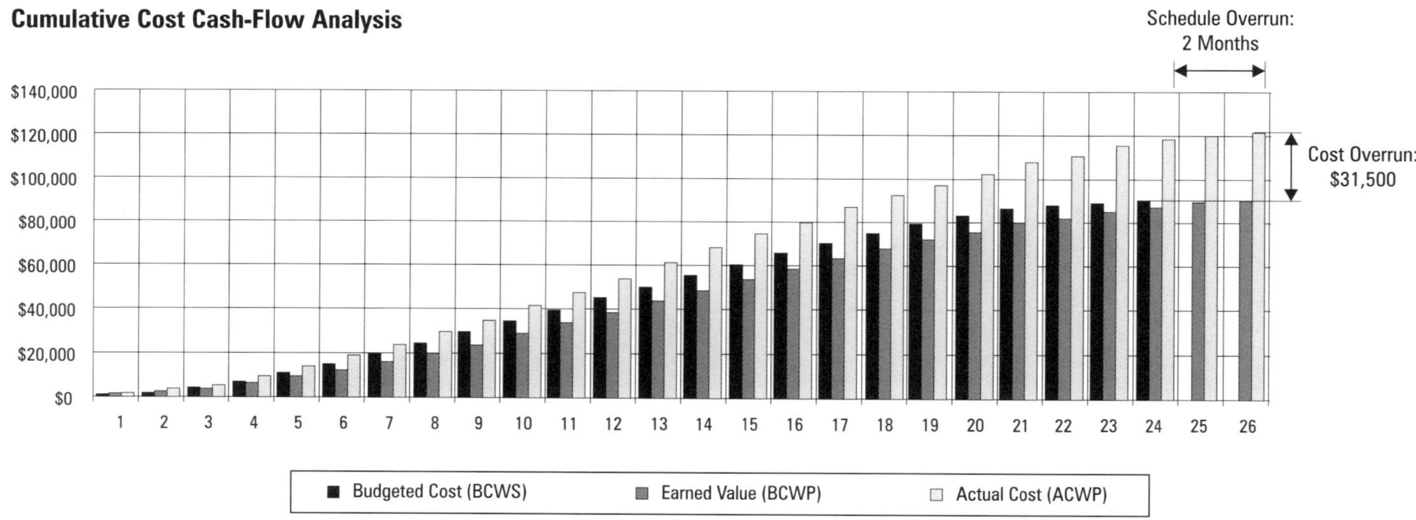

	1	2	3	4	5	6	7	8	9	10	11	12	13
Budgeted Cost (BCWS)	$1,000	$2,000	$4,000	$7,000	$11,000	$15,000	$20,000	$25,000	$30,000	$35,000	$40,000	$45,000	$50,000
Earned Value (BCWP)	$1,350	$2,700	$4,500	$6,750	$9,450	$12,600	$15,750	$19,350	$23,850	$28,800	$33,750	$38,700	$43,650
Actual Cost (ACWP)	$1,500	$3,500	$6,000	$9,500	$14,000	$18,500	$24,000	$29,500	$35,000	$41,000	$47,000	$54,000	$61,000

	14	15	16	17	18	19	20	21	22	23	24	25	26
Budgeted Cost (BCWS)	$55,000	$60,000	$65,000	$70,000	$75,000	$79,000	$83,000	$86,000	$88,000	$89,000	$90,000	$0	$0
Earned Value (BCWP)	$48,600	$53,550	$58,500	$63,450	$67,500	$71,550	$75,600	$79,200	$81,900	$84,600	$87,300	$89,100	$90,000
Actual Cost (ACWP)	$68,000	$74,000	$80,000	$86,000	$92,000	$97,000	$102,000	$107,000	$111,000	$115,000	$118,000	$120,000	$121,500

Figure 5. Continued

Continued on next page

Figure 5. Continued

Project Cost and Schedule Performance Analysis

Performance Measures

Cost Variance BCWP – ACWP	Cost Underruns or Overruns Due to Budget Performance, \$0> Is Favorable
Schedule Variance BCWP – BCWS	Cost Underruns or Overruns Due to Schedule Performance, \$0> Is Favorable
Cost Performance Index BCWP/ACWP	Ratio of Budgeted vs. Actual Cost Based on Actual Schedule, 1> Is Favorable
Schedule Performance Index BCWP/BCWS	Ratio of Budgeted Costs Based on Actual Schedule vs. Base Schedule, 1> Is Favorable

Monthly Cost and Schedule Variances

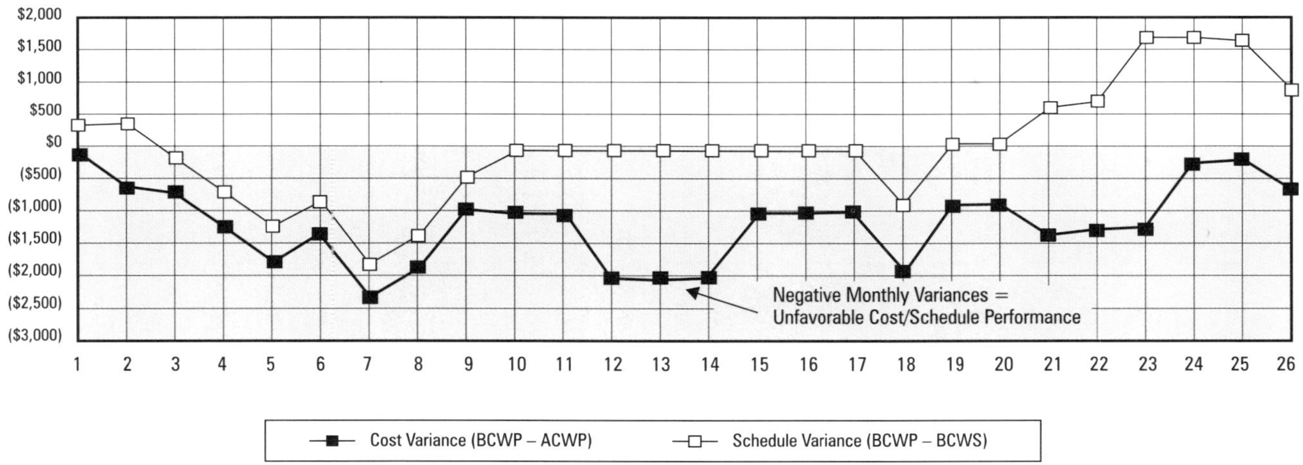

	1	2	3	4	5	6	7	8	9	10	11	12	13
Cost Variance BCWP – ACWP	(\$150)	(\$650)	(\$700)	(\$1,250)	(\$1,800)	(\$1,350)	(\$2,350)	(\$1,900)	(\$1,000)	(\$1,050)	(\$1,050)	(\$2,050)	(\$2,050)
Schedule Variance BCWP – BCWS	\$350	\$350	(\$200)	(\$750)	(\$1,300)	(\$850)	(\$1,850)	(\$1,400)	(\$500)	(\$50)	(\$50)	(\$50)	(\$50)

	14	15	16	17	18	19	20	21	22	23	24	25	26
Cost Variance BCWP – ACWP	(\$2,050)	(\$1,050)	(\$1,050)	(\$1,050)	(\$1,950)	(\$950)	(\$950)	(\$1,400)	(\$1,300)	(\$1,300)	(\$300)	(\$200)	(\$600)
Schedule Variance BCWP – BCWS	(\$50)	(\$50)	(\$50)	(\$50)	(\$950)	\$50	\$50	\$600	\$700	\$1,700	\$1,700	\$1,800	\$900

Continued on next page

Monthly Cost and Schedule Performance Indexes

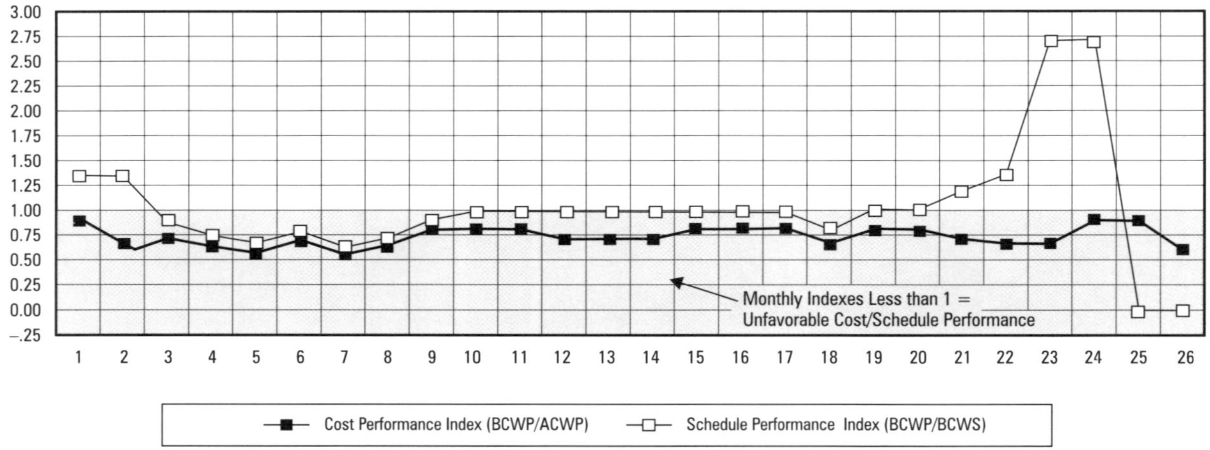

	1	2	3	4	5	6	7	8	9	10	11	12	13
Cost Performance Index BCWP/ACWP	0.90	0.68	0.72	0.64	0.60	0.70	0.57	0.65	0.82	0.83	0.83	0.71	0.71
Schedule Performance Index BCWP/BCWS	1.35	1.35	0.90	0.75	0.68	0.79	0.63	0.72	0.90	0.99	0.99	0.99	0.99

	14	15	16	17	18	19	20	21	22	23	24	25	26
Cost Performance Index BCWP/ACWP	0.71	0.83	0.83	0.83	0.68	0.81	0.81	0.72	0.68	0.68	0.90	0.90	0.60
Schedule Performance Index BCWP/BCWS	0.99	0.99	0.99	0.99	0.81	1.01	1.01	1.20	1.35	2.70	2.70		

	1	2	3	4	5	6	7	8	9	10	11	12	13
Cost Percent Overrun/(–) Underrun	-11.11%	-48.15%	-38.89%	-55.56%	-66.67%	-42.86%	-74.60%	-52.78%	-22.22%	-21.21%	-21.21%	-41.41%	-41.41%

	14	15	16	17	18	19	20	21	22	23	24	25	26
Cost Percent Overrun/(–) Underrun	-41.41%	-21.21%	-21.21%	-21.21%	-48.15%	-23.46%	-23.46%	-38.89%	-48.15%	-48.15%	-11.11%	-11.11%	-66.67%

Figure 5. Continued

Continued on next page

Cumulative Cost and Schedule Variances

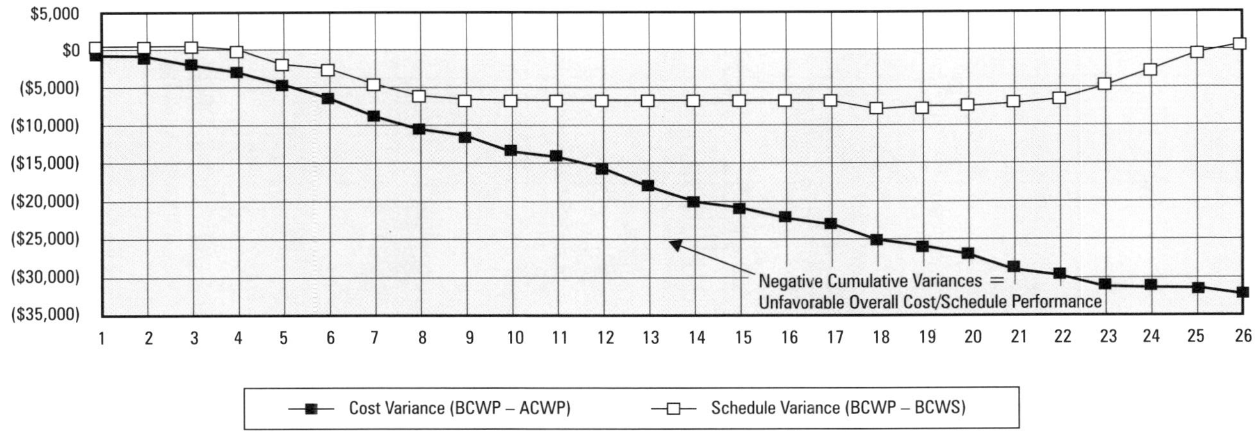

	1	2	3	4	5	6	7	8	9	10	11	12	13
Cost Variance BCWP – ACWP	($150)	($800)	($1,500)	($2,750)	($4,550)	($5,900)	($8,250)	($10,150)	($11,150)	($12,200)	($13,250)	($15,300)	($17,350)
Schedule Variance BCWP – BCWS	$350	$700	$500	($250)	($1,550)	($2,400)	($4,250)	($5,650)	($6,150)	($6,200)	($6,250)	($6,300)	($6,350)

	14	15	16	17	18	19	20	21	22	23	24	25	26
Cost Variance BCWP – ACWP	($19,400)	($20,450)	($21,500)	($22,550)	($24,500)	($25,450)	($26,400)	($27,800)	($29,100)	($30,400)	($30,700)	($30,900)	($31,500)
Schedule Variance BCWP – BCWS	($6,400)	($6,450)	($6,500)	($6,550)	($7,500)	($7,450)	($7,400)	($6,800)	($6,100)	($4,400)	($2,700)	($900)	($0)

Figure 5. Continued

Continued on next page

Cumulative Cost and Schedule Performance Indexes

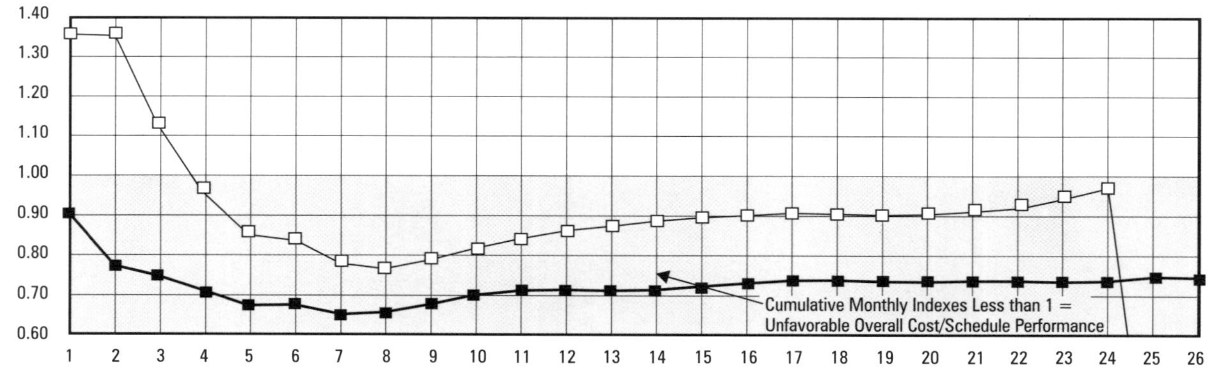

	1	2	3	4	5	6	7	8	9	10	11	12	13
Cost Performance Index BCWP/ACWP	0.90	0.77	0.75	0.71	0.68	0.68	0.66	0.66	0.68	0.70	0.72	0.72	0.72
Schedule Performance Index BCWP/BCWS	1.35	1.35	1.13	0.96	0.86	0.84	0.79	0.77	0.80	0.82	0.84	0.86	0.87

	14	15	16	17	18	19	20	21	22	23	24	25	26
Cost Performance Index BCWP/ACWP	0.71	0.72	0.73	0.74	0.73	0.74	0.74	0.74	0.74	0.74	0.74	0.74	0.74
Schedule Performance Index BCWP/BCWS	0.88	0.89	0.90	0.91	0.90	0.91	0.91	0.92	0.93	0.95	0.97		

	1	2	3	4	5	6	7	8	9	10	11	12	13
Cost Percent Overrun/(–) Underrun	-11.11%	-29.63%	-33.33%	-40.74%	-48.15%	-46.83%	-52.38%	-52.45%	-46.75%	-42.36%	-39.26%	-39.53%	-39.75%

	14	15	16	17	18	19	20	21	22	23	24	25	26
Cost Percent Overrun/(–) Underrun	-39.92%	-38.19%	-36.75%	-35.54%	-36.30%	-35.57%	-34.92%	-35.10%	-35.53%	-35.93%	-35.17%	-34.68%	-35.00%

Figure 5. Continued

Base Cost and Schedule Parameters

Total Project Cost (× 1,000)	$90,000
Project Duration (months)	24
Average Monthly Cost (× 1,000)	$3,750
Maximum Monthly Cost (× 1,000)	$5,000

Cost Measures:

Budgeted Cost (BCWS)	Budgeted Cost of Work Scheduled—Shows Base Budget Spread over Base Schedule ($90 Million in 24 Months)
Earned Value (BCWP)	Budgeted Cost of Work Performed—Shows Base Budget Spread over Actual Schedule ($90 Million in 26 Months) = BCWS × Percent Complete
Actual Cost (ACWP)	Actual Cost of Work Performed—Shows Actual Costs Expended over Actual Schedule ($121.5 Million in 26 Months)

Note: All actual costs after month 4 are projected to illustrate the recovery process.

Monthly Cost Cash-Flow Analysis

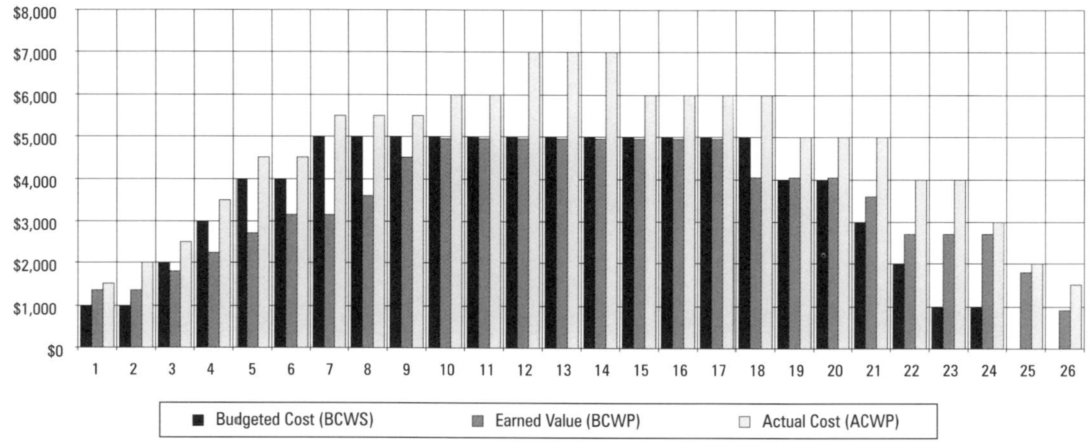

	1	2	3	4	5	6	7	8	9	10	11	12	13
Budgeted Cost (BCWS)	$1,000	$1,000	$2,000	$3,000	$4,000	$4,000	$5,000	$5,000	$5,000	$5,000	$5,000	$5,000	$5,000
Monthly Planned Percent Complete	1.11%	1.11%	2.22%	3.33%	4.44%	4.44%	5.56%	5.56%	5.56%	5.56%	5.56%	5.56%	5.56%
Monthly Actual Percent Complete	1.50%	1.50%	2.00%	2.50%	3.00%	3.50%	3.50%	4.00%	5.00%	5.50%	5.50%	5.50%	5.50%
Monthly Completion Variance	0.39%	0.39%	-0.22%	-0.83%	-1.44%	-0.94%	-2.06%	-1.56%	-0.56%	-0.06%	-0.06%	-0.06%	-0.06%
Earned Value (BCWP)	$1,350	$1,350	$1,800	$2,250	$2,700	$3,150	$3,150	$3,600	$4,500	$4,950	$4,950	$4,950	$4,950
Actual Cost (ACWP)	$1,500	$2,000	$2,500	$3,500	$4,500	$4,500	$5,500	$5,500	$5,500	$6,000	$6,000	$7,000	$7,000
Budget Variance (BCWS – ACWP)	($500)	($1,000)	($500)	($500)	($500)	($500)	($500)	($500)	($500)	($1,000)	($1,000)	($2,000)	($2,000)

	14	15	16	17	18	19	20	21	22	23	24	25	26
Budgeted Cost (BCWS)	$5,000	$5,000	$5,000	$5,000	$5,000	$4,000	$4,000	$3.000	$2,000	$1,000	$1,000	$0	$0
Monthly Planned Percent Complete	5.56%	5.56%	5.56%	5.56%	5.56%	4.44%	4.44%	3.33%	2.22%	1.11%	1.11%	0.00%	0.00%
Monthly Actual Percent Complete	5.50%	5.50%	5.50%	5.50%	4.50%	4.50%	4.50%	4.00%	3.00%	3.00%	3.00%	2.00%	1.00%
Monthly Completion Variance	-0.06%	-0.06%	-0.06%	-0.06%	-1.06%	0.06%	0.06%	0.67%	0.78%	1.89%	1.89%	2.00%	1.00%
Earned Value (BCWP)	$4,950	$4,950	$4,950	$4,950	$4,050	$4,050	$4,050	$3,600	$2,700	$2,700	$2,700	$1,800	$900
Actual Cost (ACWP)	$7,000	$6,000	$6,000	$6,000	$6,000	$5,000	$5,000	$5,000	$4,000	$4,000	$3,000	$2,000	$1,500
Budget Variance (BCWS – ACWP)	($2,000)	($1,000)	($1,000)	($1,000)	($1,000)	($1,000)	($1,000)	($2,000)	($2,000)	($3,000)	($2,000)	($2,000)	($1,500)

Figure 6. Recovery of Project over Budget and Behind Schedule—Performance/Productivity Approach

Continued on next page

Cumulative Cost Cash-Flow Analysis

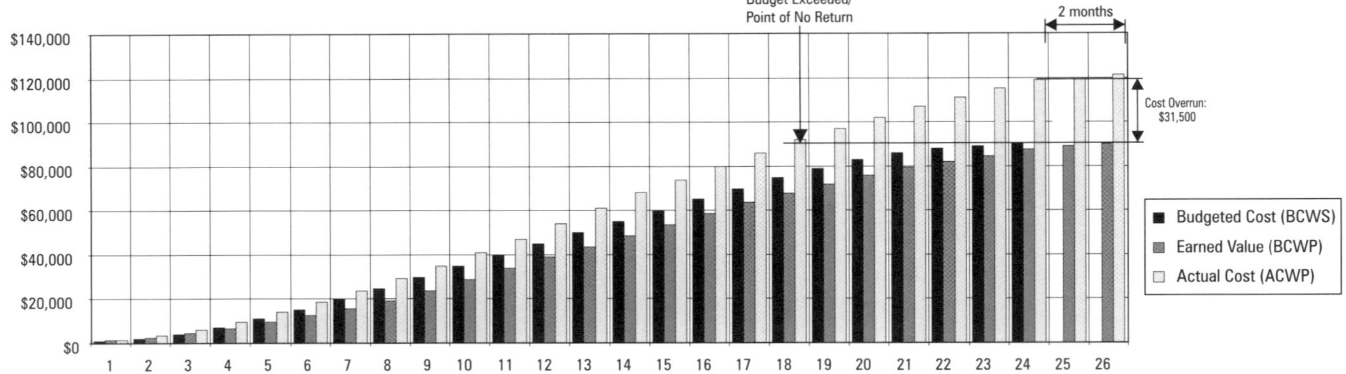

	1	2	3	4	5	6	7	8	9	10	11	12	13
Budgeted Cost (BCWS)	$1,000	$2,000	$4,000	$7,000	$11,000	$15,000	$20,000	$25,000	$30,000	$35,000	$40,000	$45,000	$50,000
Planned Percent Complete	1.11%	2.22%	4.44%	7.78%	12.22%	16.67%	22.22%	27.78%	33.33%	38.89%	44.44%	50.00%	55.56%
Actual Percent Complete	1.50%	3.00%	5.00%	7.50%	10.50%	14.00%	17.50%	21.50%	26.50%	32.00%	37.50%	43.00%	48.50%
Completion Variance Percent	0.39%	0.78%	0.56%	-0.28%	-1.72%	-2.67%	-4.72%	-6.28%	-6.83%	-6.89%	-6.94%	-7.00%	-7.06%
Schedule Recovery Requirement (months)	0.09	0.19	0.13	-0.07	-0.41	-0.64	-1.13	-1.51	-1.64	-1.65	-1.67	-1.68	-1.69
Months Remaining	23	22	21	20	19	18	17	16	15	14	13	12	11
Schedule Recovery Index (percent)	0.41%	0.85%	0.63%	-0.33%	-2.18%	-3.56%	-6.67%	-9.42%	-10.93%	-11.81%	-12.82%	-14.00%	-15.39%
Earned Value (BCWP)	$1,350	$2,700	$4,500	$6,750	$9,450	$12,600	$15,750	$19,350	$23,850	$28,800	$33,750	$38,700	$43,650
Actual Cost (ACWP)	$1,500	$3,500	$6,000	$9,500	$14,000	$18,500	$24,000	$29,500	$35,000	$41,000	$47,000	$54,000	$61,000
Budget Variance (BCWS – ACWP)	($500)	($1,500)	($2,000)	($2,500)	($3,000)	($3,500)	($4,000)	($4,500)	($5,000)	($6,000)	($7,000)	($9,000)	($11,000)
Budget Recovery Requirement	$22	$68	$95	$125	$158	$194	$235	$281	$333	$429	$538	$750	$1,000
Cost Variance BCWP – ACWP	($150)	($800)	($1,500)	($2,750)	($4,550)	($5,900)	($8,250)	($10,150)	($11,150)	($12,200)	($13,250)	($15,300)	($17,350)
Earned Value Recovery Requirement	$7	$36	$71	$138	$239	$328	$485	$634	$743	$871	$1,019	$1,275	$1,577

	14	15	16	17	18	19	20	21	22	23	24	25	26
Budgeted Cost (BCWS)	$55,000	$60,000	$65,000	$70,000	$75,000	$79,000	$83,000	$86,000	$88,000	$89,000	$90,000	$0	$0
Planned Percent Complete	61.11%	66.67%	72.22%	77.78%	83.33%	87.78%	92.22%	95.56%	97.78%	98.89%	100.00%	100.00%	100.00%
Actual Percent Complete	54.00%	59.50%	65.00%	70.50%	75.00%	79.50%	84.00%	88.00%	91.00%	94.00%	97.00%	99.00%	100.00%
Completion Variance Percent	-7.11%	-7.17%	-7.22%	-7.28%	-8.33%	-8.28%	-8.22%	-7.56%	-6.78%	-4.89%	-3.00%	-1.00%	0.00%
Schedule Recovery Requirement (months)	-1.71	-1.72	-1.73	-1.75	-2.00	-1.99	-1.97	-1.81	-1.63	-1.17	-0.72		
Months Remaining	10	9	8	7	6	5	4	3	2	1	0	-1	-2
Schedule Recovery Index (percent)	-17.07%	-19.11%	-21.67%	-24.95%	-33.33%	-39.73%	-49.33%	-60.44%	-81.33%	-117.33%			
Earned Value (BCWP)	$48,600	$53,550	$58,500	$63,450	$67,500	$71,550	$75,600	$79,200	$81,900	$84,600	$87,300	$89,100	$90,000
Actual Cost (ACWP)	$68,000	$74,000	$80,000	$86,000	$92,000	$97,000	$102,000	$107,000	$111,000	$115,000	$118,000	$120,000	$121,500
Budget Variance (BCWS – ACWP)	($13,000)	($14,000)	($15,000)	($16,000)	($17,000)	($18,000)	($19,000)	($21,000)	($23,000)	($26,000)	($28,000)	($30,000)	($31,500)
Budget Recovery Requirement	$1,300	$1,556	$1,875	$2,286	$2,833	$3,600	$4,750	$7,000	$11,500	$26,000	$28,000	$30,000	$31,500
Cost Variance BCWP – ACWP	($19,400)	($20,450)	($21,500)	($22,550)	($24,500)	($25,450)	($26,400)	($27,800)	($29,100)	($30,400)	($30,700)	($30,900)	($31,500)
Earned Value Recovery Requirement	$1,940	$2,272	$2,688	$3,221	$4,083	$5,090	$6,600	$9,267	$14,550	$30,400	$30,700	$30,900	$31,500

Figure 6. Continued

Continued on next page

Schedule Recovery Examples

Monthly Recovery Limit (months)		1	2	3	4	5	6	7	8	9	10	11	12	13	14	15	16	17	18	19	20	21	22	23	24	25	26

		1	
0.30	Start Recovery the Month Delay Occurs	0.30	Additional Labor Expended = 0.3 Months, Time to Recover 1 Month
	Delay	0.23	Schedule Recovered

		1	2	3	4	5	6	
0.30	Start Recovery 7 Months after Delay Starts	0.30	0.60	0.90	1.20	1.50	1.80	Additional Labor Expended = 1.8 Months, Time to Recover 6 Months
	Delay	(1.37)	(1.08)	(0.79)	(0.51)	(0.22)	0.07	Schedule Recovered

Budget Recovery Examples

Monthly Recovery Limit ($)		1	2	3	4	5	
$1,000	Start Recovery the Month Overrun Occurs	$1,000	$2,000	$3,000	$4,000	$5,000	Budget Recovered in 5 Months, Recovery Cost = $5,000
	Budget Overrun (Budget Variance – Recovery Amount)	($1,500)	($1,000)	($500)	$0	$500	Budget Recovered

		1	2	3	4	5	6	7	8	9	10	11	12	13	
$1,500	Start Recovery 7 Months after Overrun Starts	$1,500	$3,000	$4,500	$6,000	$7,500	$9,000	$10,500	$12,000	$13,500	$15,000	$16,500	$18,000	$19,500	Budget Recovered in 13 Months, Recovery Cost = $19,500
	Budget Overrun (Budget Variance – Recovery Amount)	($3,000)	($2,000)	($1,500)	($1,000)	($1,500)	($2,000)	($2,500)	($2,000)	($1,500)	($1,000)	($500)	$0	$500	Budget Recovered

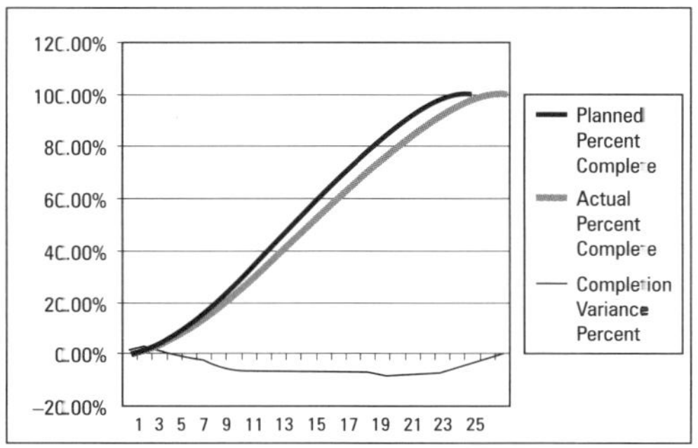

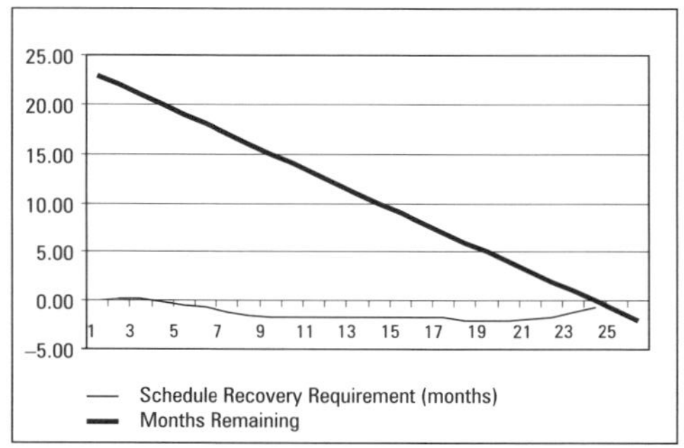

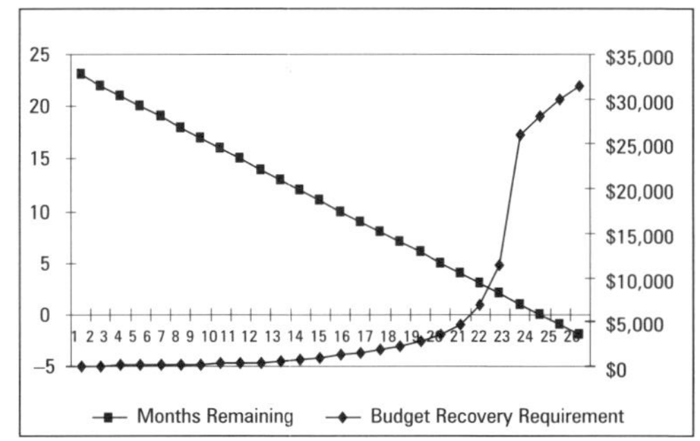

Project Cost and Schedule Performance Analysis

Performance Measures	
Cost Variance BCWP – ACWP	Cost Underruns or Overruns Due to Budget Performance, $0> Is Favorable
Schedule Variance BCWP – BCWS	Cost Underruns or Overruns Due to Schedule Performance, $0> Is Favorable
Cost Performance Index BCWP/ACWP	Ratio of Budgeted vs. Actual Cost Based on Actual Schedule, 1> Is Favorable
Schedule Performance Index BCWP/BCWS	Ratio of Budgeted Costs Based on Actual Schedule vs. Base Schedule, 1> Is Favorable

Continued on next page

Cumulative Cost and Schedule Variances

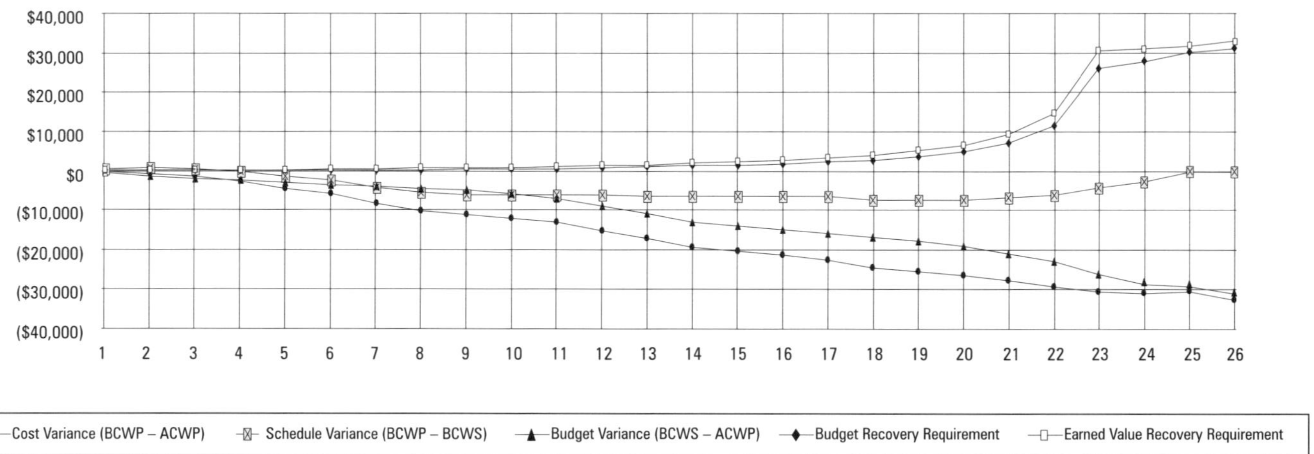

	1	2	3	4	5	6	7	8	9	10	11	12	13
Cost Variance BCWP – ACWP	($150)	($800)	($1,500)	($2,750)	($4,550)	($5,900)	($8,250)	($10,150)	($11,150)	($12,200)	($13,250)	($15,300)	($17,350)
Schedule Variance BCWP – BCWS	$350	$700	$500	($250)	($1,550)	($2,400)	($4,250)	($5,650)	($6,150)	($6,200)	($6,250)	($6,300)	($6,350)
Budget Variance (BCWS – ACWP)	($500)	($1,500)	($2,000)	($2,500)	($3,000)	($3,500)	($4,000)	($4,500)	($5,000)	($6,000)	($7,000)	($9,000)	($11,000)
Budget Recovery Requirement	$22	$68	$95	$125	$158	$194	$235	$281	$333	$429	$538	$750	$1,000
Earned Value Recovery Requirement	$7	$36	$71	$138	$239	$328	$485	$634	$743	$871	$1,019	$1,275	$1,577

	14	15	16	17	18	19	20	21	22	23	24	25	26
Cost Variance BCWP – ACWP	($19,400)	($20,450)	($21,500)	($22,550)	($24,500)	($25,450)	($26,400)	($27,800)	($29,100)	($30,400)	($30,700)	($30,900)	($31,500)
Schedule Variance BCWP – BCWS	($6,400)	($6,450)	($6,500)	($6,550)	($7,500)	($7,450)	($7,400)	($6,800)	($6,100)	($4,400)	($2,700)		
Budget Variance (BCWS – ACWP)	($13,000)	($14,000)	($15,000)	($16,000)	($17,000)	($18,000)	($19,000)	($21,000)	($23,000)	($26,000)	($28,000)	($30,000)	($31,500)
Budget Recovery Requirement	$1,300	$1,556	$1,875	$2,286	$2,833	$3,600	$4,750	$7,000	$11,500	$26,000	$28,000	$30,000	$31,500
Earned Value Recovery Requirement	$1,940	$2,272	$2,688	$3,221	$4,083	$5,090	$6,600	$9,267	$14,550	$30,400	$30,700	$30,900	$31,500

Figure 6. Continued

For example, our project recovery limit is 0.3 months, which means that we can only perform the equivalent of ten extra calendar days' worth of recovery/makeup work every month. Look at what happens if we apply the recovery in month four, as soon as the schedule recovery requirement/index becomes negative: the schedule is recovered during the same month. But if we wait until month eleven, it will take us five months to recover. Since the projected worst-case delay in this scenario is only two months, after a certain point, it does not "pay" to expend schedule recovery resources. However, in reality, the worst-case delay cannot be projected with certainty, and it is always best to implement recovery mechanisms at the first sign of a delay. Furthermore, delaying the start of recovery only exacerbates the financial losses, as we shall see.

Budget recovery requirement = budget variance/periods remaining. In this case study, we can see that the budget variance is negative from the beginning. Let us say that we implement the budget-recovery process in the first month when the budget variance is negative $1,500, and that our monthly recovery budget is only $1,000. Then it will take us five months and $5,000 to recover. If we wait until the eighth month, it will take us thirteen months to recover, and only if we increase the recovery expenditure to $1,500 a month. The sooner we start addressing our losses, the sooner we can expect to recover the budget and the less we will have to spend to achieve this goal. It sounds like common sense, and it is; however, prompt identification of budget overruns and timely implementation of recovery efforts is a rarity. On this particular project, the budget overrun causes included delayed fabrication of critical mechanical equipment, crane breakdowns, and site-access bottlenecks. In each case, biting the bullet immediately and expending the resources to fix the problem yielded the desired results. We had replaced the equipment vendor, replaced the crane and implemented a maintenance program, and reorganized site delivery logistics—all at an initial cost, but stemming further financial hemorrhage.

Crisis recovery costs money. As any project manager who has had to recover a project can attest, the corporate management usually resists having to spend money to fix a problem, because the losses are often hard to quantify and harder yet to project. Thus, problems fester, and in the end, good money gets thrown after bad.

In this case study, we have addressed budget and schedule recovery separately. On real projects, the recovery effort impacts both the schedule and budget simultaneously. We talked about the concept of recovery and salvage. Project recovery means bringing both the schedule and the budget to the original plan. It means that the cost and schedule variances are positive numbers, and that the cost and schedule indexes are above one. Very often, recovery is not possible, even with added resources.

You have explained the technical aspects of the recovery process. But what specific measures can we take to recover the cost and schedule performance?

This will vary greatly, depending on the problem, but it boils down to two major principles, which should be applied in parallel:

1. Increase the productivity of your *value-adding* activities—i.e., your earned value. Accomplish more each month than you had planned—earn more than you "deserve."

2. Reduce or eliminate your *nonvalue-adding* activities. Minimize cost to run the project ("transaction costs")—i.e., the costs of administering a project, the paper part of a project that does not directly contribute to budget, schedule, and quality performance. Eliminate duplication of roles and functional overlaps.

Here are a couple of simple performance parameters to check on a regular basis:

- Is your earned value in sync with schedule completion—e.g., at schedule midpoint, is 50 percent value of the work in place? Is cumulative cash flow on track?
- Is your overhead (cost to run the project) expended in sync with direct cost? Are you spending "management" money at a faster rate than putting work in place?

Project Recovery—Financial Approach

The second recovery approach is financial. It involves techniques required to maintain constant project solvency—i.e., positive cash balance for each fiscal period—within the constraints of funding availability, credit limits, borrowing, and financing requirements.

The productivity approach deals strictly with the internal cash flows, with the earned value of received payments. Financial approach takes into account the cash-flow receivables and payables—i.e., the cash flow of the project funding source and the cash flow of the suppliers/vendors/subcontractors.

To simplify the data, the case study for the financial approach shows how the cash-flow structure for a $5 million, twelve-month project was adjusted to minimize negative cash flow and financing requirements (see Figure 7).

Step 1 shows a project cash flow where payables to subcontractors are disbursed prior to funding from the client, requiring a high level of borrowing and high financing cost that would result in net loss to the project. Step 2 shows how the cash-flow status improves by paying our subcontractors only after getting paid by the client. In Step 3, we also hold retention on our subcontractor in excess of the retention that the client is holding on us. This further reduces our negative balance.

Steps 2 and 3 involve a lot of financial engineering, but the net improvement is still minor. We have reduced our losses from $30,000 to about $3,000 by basically rescheduling our payables to be as late as possible from our receivables.

In Step 4, we also reduce indirect payouts—i.e., reduce our overhead costs—by 10 percent and achieve a profit of over $48,000.

So, in Step 4, you are actually implementing a productivity approach to recovery.

Yes, both the productivity/performance approach and the financial approach to recovery should be implemented to yield the best results. What this case study shows is that the focus on productivity/performance approach yields better results than focusing on the financial approach.

1. SCHEDULED CASH FLOW

Total Contract $5,000,000
Retainage 10%
Interest Rate 12%

		WORK PERFORMED				PAYMENTS RECEIVED				PAYOUTS DISBURSED					PROJECT FINANCING					
PERIOD	% Complete	Indirect Costs & Profit	Direct Costs	Work in Place	Cumulative Work in Place	Billing	Retainage	Payment Received	Cumulative Payments Received	Indirect Payouts	Direct Payouts	**Total Payouts**	**Cumulative Payouts**	Cash Flow	Amount Borrowed	Amount Repaid	Cumulative Amount Borrowed	Interest	Cumulative Revolving Interest	NET PROFIT/ LOSS
1	5%	$25,000	$225,000	$250,000	$250,000	$225,000	-$25,000	$0	$0	$25,000	$225,000	**$250,000**	**$250,000**	-$250,000	$250,000	$0	-$250,000	-$2,500	-$2,500	-$252,500
2	7%	$35,000	$315,000	$350,000	$600,000	$315,000	-$35,000	$225,000	$225,000	$35,000	$315,000	**$350,000**	**$600,000**	-$125,000	$125,000	$0	-$375,000	-$1,250	-$4,050	-$379,050
3	10%	$50,000	$450,000	$500,000	$1,100,000	$450,000	-$50,000	$315,000	$540,000	$50,000	$450,000	**$500,000**	**$1,100,000**	-$185,000	$185,000	$0	-$560,000	-$1,850	-$6,386	-$566,386
4	10%	$50,000	$450,000	$500,000	$1,600,000	$450,000	-$50,000	$450,000	$990,000	$50,000	$450,000	**$500,000**	**$1,600,000**	-$50,000	$50,000	$0	-$610,000	-$500	-$7,652	-$617,652
5	10%	$50,000	$450,000	$500,000	$2,100,000	$450,000	-$50,000	$450,000	$1,440,000	$50,000	$450,000	**$500,000**	**$2,100,000**	-$50,000	$50,000	$0	-$660,000	-$500	-$9,071	-$669,071
6	10%	$50,000	$450,000	$500,000	$2,600,000	$450,000	-$50,000	$450,000	$1,890,000	$50,000	$450,000	**$500,000**	**$2,600,000**	-$50,000	$50,000	$0	-$710,000	-$500	-$10,659	-$720,659
7	10%	$50,000	$450,000	$500,000	$3,100,000	$450,000	-$50,000	$450,000	$2,340,000	$50,000	$450,000	**$500,000**	**$3,100,000**	-$50,000	$50,000	$0	-$760,000	-$500	-$12,438	-$772,438
8	10%	$50,000	$450,000	$500,000	$3,600,000	$450,000	-$50,000	$450,000	$2,790,000	$50,000	$450,000	**$500,000**	**$3,600,000**	-$50,000	$50,000	$0	-$810,000	-$500	-$14,431	-$824,431
9	10%	$50,000	$450,000	$500,000	$4,100,000	$450,000	-$50,000	$450,000	$3,240,000	$50,000	$450,000	**$500,000**	**$4,100,000**	-$50,000	$50,000	$0	-$860,000	-$500	-$16,662	-$876,662
10	8%	$40,000	$360,000	$400,000	$4,500,000	$360,000	-$40,000	$450,000	$3,690,000	$40,000	$360,000	**$400,000**	**$4,500,000**	$50,000	$0	$50,000	-$810,000	$0	-$18,662	-$828,662
11	5%	$25,000	$225,000	$250,000	$4,750,000	$225,000	-$25,000	$360,000	$4,050,000	$25,000	$225,000	**$250,000**	**$4,750,000**	$110,000	$0	$110,000	-$700,000	$0	-$20,901	-$720,901
12	5%	**$25,000**	$225,000	$250,000	$5,000,000	$225,000	-$25,000	$225,000	$4,275,000	$25,000	$225,000	**$250,000**	**$5,000,000**	-$25,000	$25,000	$0	-$725,000	-$250	-$23,660	-$748,660
13							$0	$225,000	$4,500,000	$0	$0	**$0**	**$5,000,000**	$225,000	$0	$225,000	-$500,000	$0	-$26,499	-$526,499
14	Retention Release							$500,000	$5,000,000	$0	$0	**$0**	**$5,000,000**	$500,000	$0	$500,000	$0	$0	-$29,678	-$29,678
TOTALS	100%	$500,000	$4,500,000	$5,000,000		$4,500,000	-$500,000	$5,000,000		$500,000	$4,500,000	**$5,000,000**		$0	$885,000	$885,000		-$8,350		

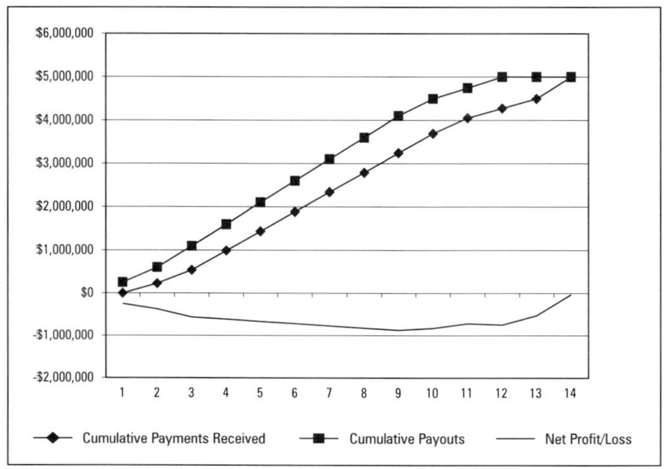

Figure 7. Project Recovery—Financial Approach

Continued on next page

2. PAY AFTER YOU HAVE BEEN PAID

Total Contract	$5,000,000
Retainage	10%
Interest Rate	12%

		WORK PERFORMED				PAYMENTS RECEIVED				PAYOUTS DISBURSED					PROJECT FINANCING					
PERIOD	% Complete	Indirect Costs & Profit	Direct Costs	Work in Place	Cumulative Work in Place	Billing	Retainage	Payment Received	Cumulative Payments Received	Indirect Payouts	Direct Payouts	**Total Payouts**	**Cumulative Payouts**	Cash Flow	Amount Borrowed	Amount Repaid	Cumulative Amount Borrowed	Interest	Cumulative Revolving Interest	NET PROFIT/ LOSS
1	5%	$25,000	$225,000	$250,000	$250,000	$225,000	-$25,000	$0	$0	$25,000	$0	**$25,000**	**$25,000**	-$25,000	$25,000	$0	-$25,000	-$250	-$250	-$25,250
2	7%	$35,000	$315,000	$350,000	$600,000	$315,000	-$35,000	$225,000	$225,000	$35,000	$225,000	**$260,000**	**$285,000**	-$35,000	$35,000	$0	-$60,000	-$350	-$630	-$60,630
3	10%	$50,000	$450,000	$500,000	$1,100,000	$450,000	-$50,000	$315,000	$540,000	$50,000	$315,000	**$365,000**	**$650,000**	-$50,000	$50,000	$0	-$110,000	-$500	-$1,206	-$111,206
4	10%	$50,000	$450,000	$500,000	$1,600,000	$450,000	-$50,000	$450,000	$990,000	$50,000	$450,000	**$500,000**	**$1,150,000**	-$50,000	$50,000	$0	-$160,000	-$500	-$1,850	-$161,850
5	10%	$50,000	$450,000	$500,000	$2,100,000	$450,000	-$50,000	$450,000	$1,440,000	$50,000	$450,000	**$500,000**	**$1,650,000**	-$50,000	$50,000	$0	-$210,000	-$500	-$2,572	-$212,572
6	10%	$50,000	$450,000	$500,000	$2,600,000	$450,000	-$50,000	$450,000	$1,890,000	$50,000	$450,000	**$500,000**	**$2,150,000**	-$50,000	$50,000	$0	-$260,000	-$500	-$3,381	-$263,381
7	10%	$50,000	$450,000	$500,000	$3,100,000	$450,000	-$50,000	$450,000	$2,340,000	$50,000	$450,000	**$500,000**	**$2,650,000**	-$50,000	$50,000	$0	-$310,000	-$500	-$4,287	-$314,287
8	10%	$50,000	$450,000	$500,000	$3,600,000	$450,000	-$50,000	$450,000	$2,790,000	$50,000	$450,000	**$500,000**	**$3,150,000**	-$50,000	$50,000	$0	-$360,000	-$500	-$5,301	-$365,301
9	10%	$50,000	$450,000	$500,000	$4,100,000	$450,000	-$50,000	$450,000	$3,240,000	$50,000	$450,000	**$500,000**	**$3,650,000**	-$50,000	$50,000	$0	-$410,000	-$500	-$6,437	-$416,437
10	8%	$40,000	$360,000	$400,000	$4,500,000	$360,000	-$40,000	$450,000	$3,690,000	$40,000	$450,000	**$490,000**	**$4,140,000**	-$40,000	$40,000	$0	-$450,000	-$400	-$7,610	-$457,610
11	5%	$25,000	$225,000	$250,000	$4,750,000	$225,000	-$25,000	$360,000	$4,050,000	$25,000	$360,000	**$385,000**	**$4,525,000**	-$25,000	$25,000	$0	-$475,000	-$250	-$8,773	-$483,773
12	5%	$25,000	$225,000	$250,000	$5,000,000	$225,000	-$25,000	$225,000	$4,275,000	$25,000	$225,000	**$250,000**	**$4,775,000**	-$25,000	$25,000	$0	-$500,000	-$250	-$10,076	-$510,076
13							$0	$225,000	$4,500,000	$0	$225,000	**$225,000**	**$5,000,000**	$0	$0	$0	-$500,000	$0	-$11,285	-$511,285
14	Retention Release							$500,000	$5,000,000	$0	$0	**$0**	**$5,000,000**	$500,000	$0	$500,000	$0	$0	-$12,639	**-$12,639**
TOTALS	100%	$500,000	$4,500,000	$5,000,000		$4,500,000	-$500,000	$5,000,000		$500,000	$4,500,000	**$5,000,000**		$0	$500,000	$500,000		-$5,000		

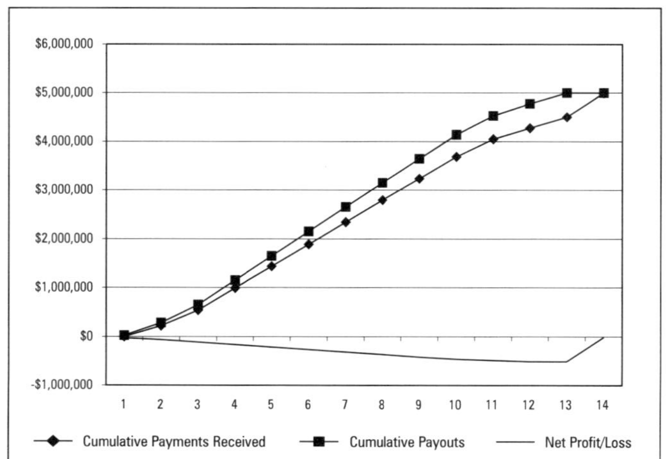

Figure 7. Continued

Continued on next page

3. INCREASE RETAINAGE ON PAYOUTS

Total Contract $5,000,000
Retainage 10%
Interest Rate 12%

		WORK PERFORMED				PAYMENTS RECEIVED				PAYOUTS DISBURSED					PROJECT FINANCING					
PERIOD	% Complete	Indirect Costs & Profit	Direct Costs	Work in Place	Cumulative Work in Place	Billing	Retainage	Payment Received	Cumulative Payments Received	Indirect Payouts	Direct Payouts	**Total Payouts**	**Cumulative Payouts**	Cash Flow	Amount Borrowed	Amount Repaid	Cumulative Amount Borrowed	Interest	Cumulative Revolving Interest	NET PROFIT/ LOSS
1	5%	$25,000	$225,000	$250,000	$250,000	$225,000	-$25,000	$0	$0	$25,000	$0	**$25,000**	**$25,000**	-$25,000	$25,000	$0	-$25,000	-$250	-$250	-$25,250
2	7%	$35,000	$315,000	$350,000	$600,000	$315,000	-$35,000	$225,000	$225,000	$35,000	$202,500	**$237,500**	**$262,500**	-$12,500	$12,500	$0	-$37,500	-$125	-$405	-$37,905
3	10%	$50,000	$450,000	$500,000	$1,100,000	$450,000	-$50,000	$315,000	$540,000	$50,000	$283,500	**$333,500**	**$596,000**	-$18,500	$18,500	$0	-$56,000	-$185	-$639	-$56,639
4	10%	$50,000	$450,000	$500,000	$1,600,000	$450,000	-$50,000	$450,000	$990,000	$50,000	$405,000	**$455,000**	**$1,051,000**	-$5,000	$5,000	$0	-$61,000	-$50	-$765	-$61,765
5	10%	$50,000	$450,000	$500,000	$2,100,000	$450,000	-$50,000	$450,000	$1,440,000	$50,000	$405,000	**$455,000**	**$1,506,000**	-$5,000	$5,000	$0	-$66,000	-$50	-$907	-$66,907
6	10%	$50,000	$450,000	$500,000	$2,600,000	$450,000	-$50,000	$450,000	$1,890,000	$50,000	$405,000	**$455,000**	**$1,961,000**	-$5,000	$5,000	$0	-$71,000	-$50	-$1,066	-$72,066
7	10%	$50,000	$450,000	$500,000	$3,100,000	$450,000	-$50,000	$450,000	$2,340,000	$50,000	$405,000	**$455,000**	**$2,416,000**	-$5,000	$5,000	$0	-$76,000	-$50	-$1,244	-$77,244
8	10%	$50,000	$450,000	$500,000	$3,600,000	$450,000	-$50,000	$450,000	$2,790,000	$50,000	$405,000	**$455,000**	**$2,871,000**	-$5,000	$5,000	$0	-$81,000	-$50	-$1,443	-$82,443
9	10%	$50,000	$450,000	$500,000	$4,100,000	$450,000	-$50,000	$450,000	$3,240,000	$50,000	$405,000	**$455,000**	**$3,326,000**	-$5,000	$5,000	$0	-$86,000	-$50	-$1,666	-$87,666
10	8%	$40,000	$360,000	$400,000	$4,500,000	$360,000	-$40,000	$450,000	$3,690,000	$40,000	$405,000	**$445,000**	**$3,771,000**	$5,000	$0	$5,000	-$81,000	$0	-$1,866	-$82,866
11	5%	$25,000	$225,000	$250,000	$4,750,000	$225,000	-$25,000	$360,000	$4,050,000	$25,000	$324,000	**$349,000**	**$4,120,000**	$11,000	$0	$11,000	-$70,000	$0	-$2,090	-$72,090
12	5%	$25,000	$225,000	$250,000	$5,000,000	$225,000	-$25,000	$225,000	$4,275,000	$25,000	$202,500	**$227,500**	**$4,347,500**	-$2,500	$2,500	$0	-$72,500	-$25	-$2,366	-$74,866
13							$0	$225,000	$4,500,000	$0	$202,500	**$202,500**	**$4,550,000**	$22,500	$0	$22,500	-$50,000	$0	-$2,650	-$52,650
14	Retention Release							$500,000	$5,000,000	$0	$450,000	**$450,000**	**$5,000,000**	$50,000	$0	$50,000	$0	$0	-$2,968	**-$2,968**
TOTALS	100%	$500,000	$4,500,000	$5,000,000		$4,500,000	-$500,000	$5,000,000		$500,000	$4,500,000	**$4,550,000**		$0	$88,500	$88,500		-$885		

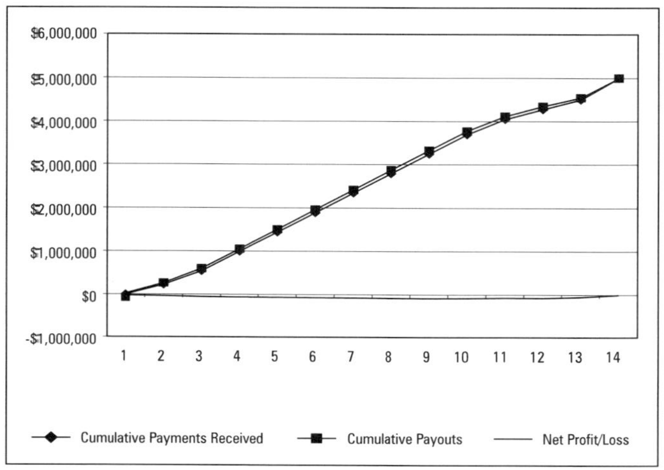

Figure 7. Continued

Continued on next page

4. REDUCE INDIRECT COST

Total Contract $5,000,000
Retainage 10%
Interest Rate 12%

		WORK PERFORMED				PAYMENTS RECEIVED				PAYOUTS DISBURSED					PROJECT FINANCING					
PERIOD	% Complete	Indirect Costs & Profit	Direct Costs	Work in Place	Cumulative Work in Place	Billing	Retainage	Payment Received	Cumulative Payments Received	Indirect Payouts	Direct Payouts	**Total Payouts**	**Cumulative Payouts**	Cash Flow	Amount Borrowed	Amount Repaid	Cumulative Amount Borrowed	Interest	Cumulative Revolving Interest	NET PROFIT/ LOSS
1	5%	$25,000	$225,000	$250,000	$250,000	$225,000	-$25,000	$0	$0	$22,500	$0	**$22,500**	**$22,500**	-$22,500	$22,500	$0	-$22,500	-$225	-$225	-$22,725
2	7%	$35,000	$315,000	$350,000	$600,000	$315,000	-$35,000	$225,000	$225,000	$31,500	$202,500	**$234,000**	**$256,500**	-$9,000	$9,000	$0	-$31,500	-$90	-$342	-$31,842
3	10%	$50,000	$450,000	$500,000	$1,100,000	$450,000	-$50,000	$315,000	$540,000	$45,000	$283,500	**$328,500**	**$585,000**	-$13,500	$13,500	$0	-$45,000	-$135	-$518	-$45,518
4	10%	$50,000	$450,000	$500,000	$1,600,000	$450,000	-$50,000	$450,000	$990,000	$45,000	$405,000	**$450,000**	**$1,035,000**	$0	$0	$0	-$45,000	$0	-$580	-$45,580
5	10%	$50,000	$450,000	$500,000	$2,100,000	$450,000	-$50,000	$450,000	$1,440,000	$45,000	$405,000	**$450,000**	**$1,485,000**	$0	$0	$0	-$45,000	$0	-$650	-$45,650
6	10%	$50,000	$450,000	$500,000	$2,600,000	$450,000	-$50,000	$450,000	$1,890,000	$45,000	$405,000	**$450,000**	**$1,935,000**	$0	$0	$0	-$45,000	$0	-$728	-$45,728
7	10%	$50,000	$450,000	$500,000	$3,100,000	$450,000	-$50,000	$450,000	$2,340,000	$45,000	$405,000	**$450,000**	**$2,385,000**	$0	$0	$0	-$45,000	$0	-$815	-$45,815
8	10%	$50,000	$450,000	$500,000	$3,600,000	$450,000	-$50,000	$450,000	$2,790,000	$45,000	$405,000	**$450,000**	**$2,835,000**	$0	$0	$0	-$45,000	$0	-$913	-$45,913
9	10%	$50,000	$450,000	$500,000	$4,100,000	$450,000	-$50,000	$450,000	$3,240,000	$45,000	$405,000	**$450,000**	**$3,285,000**	$0	$0	$0	-$45,000	$0	-$1,023	-$46,023
10	8%	$40,000	$360,000	$400,000	$4,500,000	$360,000	-$40,000	$450,000	$3,690,000	$36,000	$405,000	**$441,000**	**$3,726,000**	$9,000	$0	$9,000	-$36,000	$0	-$1,145	-$37,145
11	5%	$25,000	$225,000	$250,000	$4,750,000	$225,000	-$25,000	$360,000	$4,050,000	$22,500	$324,000	**$346,500**	**$4,072,500**	$13,500	$0	$13,500	-$22,500	$0	-$1,283	-$23,783
12	5%	$25,000	$225,000	$250,000	$5,000,000	$225,000	-$25,000	$225,000	$4,275,000	$22,500	$202,500	**$225,000**	**$4,297,500**	$0	$0	$0	-$22,500	$0	-$1,437	-$23,937
13							$0	$225,000	$4,500,000	$0	$202,500	**$202,500**	**$4,500,000**	$22,500	$0	$22,500	$0	$0	-$1,609	-$1,609
14	Retention Release							$500,000	$5,000,000	$0	$450,000	**$450,000**	**$4,950,000**	$50,000	$0	$50,000	$50,000	$0	-$1,802	**$48,198**
TOTALS	100%	$500,000	$4,500,000	$5,000,000		$4,500,000	-$500,000	$5,000,000		$450,000	$4,500,000	**$4,950,000**		$50,000	$45,000	$95,000		-$450		

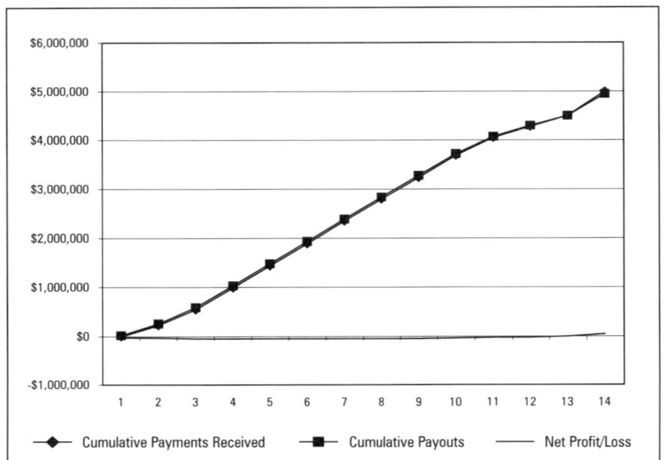

Figure 7. Continued

Conclusion

In the end, by carefully monitoring the performance indicators on a regular basis, we were able to pinpoint the budget and schedule problems before they got out of hand, and implement recovery measures immediately. The project was completed on schedule and budget.

Triage

Triage is a process of logical allocation of limited resources in order to achieve the maximum result. It is a crucial part of the emergency-management process, and its purpose is to provide an optimum balance between two objectives.

Objective 1: Fix those problems that have the best chance of survival (salvageability).

To be balanced with:

Objective 2: Fix those problems that have the best salvage value.

Focus of the triage procedure is on allocating limited resources, especially time, to save the maximum number of casualties. Just as the medical emergencies must be completed during the "golden hour"—the time during which the casualties have to be evacuated from the scene of the accident and taken to a hospital emergency room—time is of the essence in business emergencies.

We have previously discussed the difference between project recovery and salvage. In triage, we will assume that the crisis has progressed beyond the point of no return, and that our objective is to salvage the project to the best extent possible.

Imagine that you are an army field medic evacuating battlefield casualties. You have thirty wounded men, but not the resources (time, manpower, medication) to save all. Ten are mortally wounded and will probably not survive. Ten have wounds that are serious but survivable. Ten are only lightly wounded. Your objective is to save as many people as possible. If untreated, a total of seventeen people will die.

Who do you save, in which order, and how do you allocate your resources? What are your options? You can apply the same care to everyone, treat the worst cases first, or treat the least serious cases first.

Apply the Same Care to Everyone

This is probably your first instinctive reaction—save everyone in the order in which they are brought—i.e., first come, first served. However, you will spread yourself too thin; everyone is going to get the same minimal care, maybe only a field dressing and a shot of morphine. Ten that are mortally wounded will die soon anyway. Many of those less seriously injured that could have been saved with priority care will die while waiting—say, five. The "walking" wounded will probably survive with or without the immediate care. Score: 30 treated, 15 dead, or 50 percent total and treated mortality.

Treat the Worst Cases First

Say that two of the ten worst cases survive because of immediate care, but six in the "medium" group and one in the "light" group die because of the lack of it. Score: 30 treated, 15 dead, or 50 percent total and treated mortality.

Treat the Least Serious Cases First

Again, these ten will probably survive without emergency care while, in the meantime, some casualties with more serious but survivable injuries will die unnecessarily due to delayed care—say, five. By the time you get to the mortally wounded, they will be almost certainly dead. Score: 30 treated, 15 dead.

Note that in the three cases, all efforts yield essentially the same result, even though they differ vastly in approach. Even with the enormous effort expended, the survival rate is only marginally better than if you had never shown up in the first place.

The correct answer is "none of the above"—*you must treat the survivable group first*. The least serious will probably survive anyway. The mortals will probably die. Score: 10 treated, 20 survived.

While this is a simplistic and extreme example, the concept is valid in emergency situations when hard decisions have to be made quickly and the effect of resources maximized. Note that, as used in this book, the term *survivability* is used interchangeably with the term *salvageability*, and the term *survival value* is used interchangeably with the term *salvage value*.

In order to maximize the survivability (salvageability) and survival value (salvage value; more about this later) you must:

1. **Prioritize—Properly "tag" the casualties: treat/don't treat and in which order.**
2. **Expedite—Ensure that treatment is applied promptly.**

✔ Save those that have the best chance of survival (survivability = salvageability),

and, although this is seldom admitted in human casualty terms:

✔ Save those who are the most deserving of survival (survival value = salvage value).

Let us illustrate the survival-value concept. If I have five men with equally serious injuries—four who are ordinary riflemen, and only one is a crack sharpshooter—I will focus my effort on saving the sharpshooter, simply because he is: 1) more valuable to me alive than dead, and 2) more valuable to me than the other four.

Now, replace "medic" with "manager" and "casualties" with "business problems." The battlefield or disaster scene triage principles apply equally to the corporate crisis survival objectives, as illustrated in the case study at the end of this triage discussion, A Case in Point—Project Triage.

Another principle of battlefield casualty management applicable to business crisis situations is the so-called *mass casualty situation*. In a troubled company or on a troubled project, when it rains, it pours. Problems will not appear in an orderly, logical sequence, but rather as a deluge of seemingly unrelated crisis events, each clamoring for the troubleshooter's attention. Attempting to address all incoming problems at the same time or on a first-come, first-served basis will lead to confusion, backlogs, and bottlenecks, and will only exacerbate a bad situation. You must learn how to quickly evaluate and sort the problems,

Priority by Salvage Value	
Salvageability	Salvage Value
A	1
B	1
C	1
A	2
B	2
C	2

Priority by Salvageability	
Salvageability	Salvage Value
A	1
A	2
B	1
B	2
C	1
C	2

A=Salvageable B=Minor C=Terminal
1=Important 2=Not Important

Table 2. Application of Triage to Business Problems

based on their relative importance and survivability, and prioritize the allocation of your resources to ensure the survival of the most and the fittest.

By applying triage methods, the receipt and disposition of a large number of simultaneous problems becomes manageable and minimizcs confusion and panic of a random first-come, first-served approach. With proper triage, available resources are applied to the maximum number of salvageable problems. Simple lifesaving procedures, which can be rapidly performed, should be given the highest priority. As in war, life takes precedence over limb, and functional repair over cosmetic appearance. The intent is to seal off the damage and buy time.

Some problems are close to having a tag on the toe by the time they are uncovered. Military terminology calls these "expectant casualties," as in "expected to die." Expectant problems are disastrous to the point that even if they were the only problem manifested and had all available emergency resources at their disposal, their chances of salvage would still be nil. During a "mass-casualty" situation, a problem beyond repair, no matter how important it is, would require unjustified commitment of limited resources, while in the meantime, salvageable problems wait and deteriorate.

Table 2 summarizes the application of triage to business problems. The first approach allocates priority by salvage value, the second, by salvageability. Problems have to be categorized accordingly; for example, A1 is a salvageable and an important problem; problem B2 is minor and not important; and so on.

Conclusion

Problem-solving resources must always be applied to salvageable problems first. The expectant problems must be written off. You are practicing economic Darwinism—survival of the fittest—with all its implications and consequences.

Do not treat those that:

✔ Will survive anyway.

✔ Will die anyway.

However, hand in hand with cutting your losses, you must also develop a habit of supporting your winners. A study of military history shows that major battles are won not by shoring up failing defenses, by dissipating the back-up resources, but by providing overwhelming, concentrated support to the units that are making a breakthrough in the enemy lines.

Start starving your problems and feeding your opportunities now! Remember:

- Concentration of Resources = Victory.
- Dispersal of Resources = Defeat.

A Case in Point—Project Triage

I will use a public transportation agency project to illustrate the application of triage in a project environment. This project had serious cost, schedule, and quality problems, including the bankruptcy of three major subcontractors, failure to procure critical components, undocumented and unpaid changes, and subcontractors' claims amounting to 5 percent of the project value—among others. The Project Surgeon's involvement started about halfway into the two-year project, and with major reorganization of the project control and procurement systems, the project was completed at 8 percent net profit. This was not bad for a public bid during the construction industry slump in the early 1990s. I have changed the problem examples and made the cost issues more severe in order to illustrate the drastic choices that have to be made under pressure. However, the thought process and triage methodology is the same that the Project Surgeon used on the actual project.

Problems on a project appear at random: some are urgent; others can be deferred until tomorrow or ignored altogether. A major issue often follows a minor nuisance. First and foremost, the project manager must keep an ongoing list of problem issues—I call this a *triage list*. Not all problems should make it on the list. Some are minor and can be ignored; others, even if serious, can be resolved quickly. Triage process in these cases can be applied quickly, using our common sense and project management experience. The problems that make it to the triage list are those that require analysis to determine if resources—time, money, manpower—should be expended to resolve them. The project in our case study was typical in that the problems occurred at random. However, the project team attempted to resolve the problems in a similarly random fashion. Every time a new problem appeared, the previous one was dropped, and so on. When things got slow, someone would mention an old issue: "Whatever happened to? ... " Since nobody kept a triage list, there was no method to identify and address or bring closure to the problems—if they got resolved, fine; if not, they festered. When I got involved, the project was already a year old, and needless to say, some of the problems had chronically lingered for a year.

The first thing the Project Surgeon did was to develop a preliminary triage list, going back to the beginning of the project. This was nothing more than a handwritten checklist on a yellow legal pad. The list was pages long—a year's worth of problems, real or imagined, jostling for space on the notepad. The second step was to cross out all the "cold cases": items that were obviously obsolete, minor, or otherwise irrelevant to the recovery of the project.

Next, the Project Surgeon listed the remaining real problems in a triage matrix, reproduced in Figure 8 (see page 53). A triage matrix is an expanded triage list; in addition to listing the problems, it also identifies the main criteria on which the problems will be triaged for resolution. A triage matrix must be kept current and short. The intent is to triage the listed problems on a set regular schedule, daily or weekly, and dispose of each problem on the list, so that room can be made for the inevitable new problems. Our capacity/resources to handle the problems must be balanced with the volume of problems. Missing a few regular triage runs will overwhelm our resolution resources. The frequency of triage will vary, depending on the circumstances of each project, the volume and gravity of the problems, and your project management systems. I prefer to dispose of the majority of problems on the triage list at the end of the week; the problems may be treated in the recovery stage, deferred until the salvage stage, or written off. The main reason for a weekly "disposal" is that my project management system issues most status reports on a weekly basis—schedule delays, production status, etc.—allowing me to compare the problem parameter to the most current data. Actually, the main benefit is that you can go home for the weekend and start with a clean slate on Monday. Disposal of problems by means of triage does not mean that all problems will be resolved; it only means that all problems were identified, assessed, and acted upon.

If we had all the resources at our disposal—including unlimited time, money, and manpower to attack and resolve all the problems on the triage list—we would not need the triage procedures. However, this is almost never the case. We usually only have the resources to address a limited number of problems. The knee-jerk instinctive reaction is to attempt to treat all the problems. The only certain outcome of this random approach is that our resources will be expended with only minor recovery effect to show.

In our case study, we have a list of ten problems, with a total estimated recovery cost of $1,445,000. If we had this money to burn, and if there was no time limit on action, we could attempt to solve the problems at our leisure in any random order. A lot of companies tend to dispose of their corporate and project disasters in this manner: writing off losses or taking a "one-time charge" on their income statements. The tendency is to make up for current losses by demanding ever-higher profits on future projects instead of treating each project as a separate profit/loss center. Very few companies can afford to do this for a long time. Therefore, for the purposes of the case study, assume that we can only address five problems on the list. Depending on your specific project conditions, you can also set a recovery budget limit—say, $500,000 in this case.

Our triage matrix example provides for addressing the problems at the breakeven point (recovery) or at the point of no return (salvage). The reason for this is twofold. First, not all problems that cannot or should not be fully recovered are hopelessly lost; often the salvage rate remains high, and the salvage cost is low—even lower than the recovery cost. Second, the final loss for some problems actually decreases over time, between the recovery and salvage phases. In both cases, we can defer resolution until the salvage phase and save recovery resources for more critical problems.

Running Triage at the Breakeven Point

By definition, if action is implemented at or before the breakeven point, we expect a full recovery and no losses. Therefore, all triage in the recovery stage is performed on the basis of importance criteria and not on salvageability. The objective of triage at the breakeven point is to minimize the total of the recovery cost and unrecovered loss.

However, it is important to project the salvageable percentage that can be expected if we defer triage until the point of no return. For example, if the problem will be only 10 percent salvageable in three months and it is also very important, we may want to recover immediately, even though the recovery cost is high. If the problem is not very important, or the salvage cost is excessive, we may want to sacrifice it. The lower the projected salvageability and the higher the importance, the more priority that should be placed on recovery.

Now, let us look at the triage matrix in detail (see Figure 8):

- Column A lists problems in the order of appearance. When we perform the triage, these numbers will be listed in order of precedence.
- Column B is the Description of the Problem.
- Column C shows the Potential Loss directly related to the Problem Activity.
- Column D shows the total Potential Impact Loss on the entire project resulting from a particular Problem. For example, the potential loss for Problem 7 alone is $2,000, but if unrecovered, it could impact a number of future activities and grow into a $500,000 loss.
- Column E is the Loss Leverage factor or Potential Impact Loss divided by Potential Problem Loss. The bigger this factor, the more damage a Problem can cause in the long run, if unaddressed.
- Column F is the Recovery Cost. It shows how much money must be expended to recover the Problem.
- Column G is the Recovery Leverage factor or Recovery Cost divided by Potential Impact Loss. It shows the bang for our recovery buck: the lower, the better.
- Column H is the Recovery Deadline date. This is the breakeven point date for the problem, the last chance to fully recover.
- Column I is the remaining time to recover, in days.

Columns D through I represent the criteria for running triage at the breakeven point. As discussed, since the objective of triage at the breakeven point is full recovery, the only triage criterion, therefore, is importance (criticality). Triage is run on the following six steps to determine the order in which the problems should be addressed. Depending on the total number of problems and the number of problems with expiring breakeven points, it may be advisable to sort by Column I first and then run secondary triage sorts on the other four cost factors. In our case, we will assume that sufficient time remains for all problems to be triaged on schedule, so that we will not presort on time.

- Step 1: Column I, minimum remaining time to recover (presort if applicable).
- Step 2: Column D, maximum Potential Impact Loss (see Figure 9).
- Step 3: Column E, maximum Loss Leverage Factor (see Figure 10, page 57).
- Step 4: Column F, minimum Recovery Cost (see Figure 11, page 58).
- Step 5: Column G, minimum Recovery Leverage factor (see Figure 12, page 61).
- Step 6: Check the minimum total Recovery Cost plus unrecovered loss for each triage sort, and choose the overall minimum. Address the problems in this order.

The best triage results will vary, depending on variations or relative differences in numbers within one problem or among all problems triaged. For some runs, best results will be achieved with a sort on maximum potential impact loss; for others, on minimum recovery cost.

Limits: Only Top Five Problems Can Be Addressed

Recovery Phase—Breakeven Point

Current Date	9/22/00

A	B	C	D	E=D/C	F	G=F/D	H
Problem	Description	Problem Potential Loss	Potential Impact Loss	Loss Leverage Factor	Recovery Cost	Recovery Leverage	Recovery Deadline
9	Rigging Equipment Breakdowns	$11,000	$875,000	79.55	$300,000	0.34	9/30/00
8	Electrical Contract Scope Errors	$15,000	$785,000	52.33	$75,000	0.10	9/24/00
10	Incomplete Survey of Existing Conditions	$2,000	$775,000	387.50	$150,000	0.19	9/27/00
7	Unsubmitted Change Orders	$2,000	$500,000	250.00	$50,000	0.10	10/2/00
6	Delayed Control System Approvals	$30,000	$450,000	15.00	$575,000	1.28	9/23/00
4	Steel Support Misfabrication	$4,000	$200,000	50.00	$195,000	0.98	9/25/00
1	Lost Fabrication Drawings for Air Handlers	$25,000	$175,000	7.00	$30,000	0.17	10/1/00
3	Leak in Tank Vessel	$7,000	$47,000	6.71	$15,000	0.32	10/5/00
5	Switchgear Malfunctions	$25,000	$25,000	1.00	$15,000	0.60	9/24/00
2	Communication System Malfunction	$5,000	$15,000	3.00	$40,000	2.67	9/28/00
Totals		**$126,000**	**$3,847,000**		**$1,445,000**		
Recovery or Salvage		**$60,000**	**$3,385,000**		**$1,150,000**	**0.34**	
Unrecovered/Unsalvaged Loss			**$462,000**				
Total Recovery/Salvage Cost +Unrecovered Loss			**$1,612,000**				

Cumulative Impact Loss vs. Recovery Cost Analysis					
9	Rigging Equipment Breakdowns	$875,000	$300,000	0.34	
8	Electrical Contract Scope Errors	$1,660,000	$375,000	0.23	
10	Incomplete Survey of Existing Conditions	$2,435,000	$525,000	0.22	
7	Unsubmitted Change Orders	$2,935,000	$575,000	0.20	min leverage
6	Delayed Control System Approvals	$3,385,000	$1,150,000	0.34	
4	Steel Support Misfabrication	$3,585,000	$1,345,000	0.38	
1	Lost Fabrication Drawings for Air Handlers	$3,760,000	$1,375,000	0.37	
3	Leak In Tank Vessel	$3,807,000	$1,390,000	0.37	
5	Switchgear Malfunctions	$3,832,000	$1,405,000	0.37	
2	Communication System Malfunction	$3,847,000	$1,445,000	0.38	

Figure 9. Triage on Maximum Potential Impact Loss

Salvage Phase—Point of No Return

	J	K	L	M=J+L	N=M/D	O=M/F	P	Q	R=P–I
naining over	Salvage Cost	Salvageability	Final Projected Loss	Loss Including Salvage Cost	Salvage Leverage	Salvage/ Recovery Cost Premium	Salvage Deadline	Days Remaining to Salvage	Breakeven/No Return Lag
	$50,000	20%	$140,000	$190,000	1.09	6.33	10/15/00	23	14
	$75,000	30%	$10,500	$85,500	5.70	2.14	9/30/00	8	2
	$35,000	50%	$23,500	$58,500	1.24	3.90	11/20/00	59	46
	$225,000	60%	$80,000	$305,000	1.53	1.56	12/2/00	71	68
	$50,000	20%	$20,000	$70,000	2.80	4.67	10/12/00	20	18
	$875,000	30%	$315,000	$1,190,000	2.64	2.07	10/15/00	23	22
	$35,000	10%	$450,000	$485,000	0.97	9.70	10/25/00	33	23
	$150,000	35%	$510,250	$660,250	0.84	8.80	12/5/00	74	72
	$350,000	60%	$350,000	$700,000	0.80	2.33	11/7/00	46	38
	$500,000	10%	$697,500	$1,197,500	1.55	7.98	9/30/00	8	3
	$2,345,000		**$2,596,750**	**$4,941,750**					
	$435,000		**$274,000**	**$709,000**	**1.53**	**2.40**			
				$4,232,750					

point of diminishing returns

Problems Listed in Order of Appearance

Limits: Only Top Five Problems Can Be Addressed

Recovery Phase—Breakeven Point

Current Date 9/22/00

A	B	C	D	E=D/C	F	G=F/D	H	
Problem	Description	Problem Potential Loss	Potential Impact Loss	Loss Leverage Factor	Recovery Cost	Recovery Leverage	Recovery Deadline	Days Re… to Rec…
1	Lost Fabrication Drawings for Air Handlers	$25,000	$175,000	7.00	$30,000	0.17	10/1/00	9
2	Communication System Malfunction	$5,000	$15,000	3.00	$40,000	2.67	9/28/00	6
3	Leak in Tank Vessel	$7,000	$47,000	6.71	$15,000	0.32	10/5/00	1[illegible]
4	Steel Support Misfabrication	$4,000	$200,000	50.00	$195,000	0.98	9/25/00	3
5	Switchgear Malfunctions	$25,000	$25,000	1.00	$15,000	0.60	9/24/00	2
6	Delayed Control System Approvals	$30,000	$450,000	15.00	$575,000	1.28	9/23/00	1
7	Unsubmitted Change Orders	$2,000	$500,000	250.00	$50,000	0.10	10/2/00	1[illegible]
8	Electrical Contract Scope Errors	$15,000	$785,000	52.33	$75,000	0.10	9/24/00	2
9	Rigging Equipment Breakdowns	$11,000	$875,000	79.55	$300,000	0.34	9/30/00	8
10	Incomplete Survey of Existing Conditions	$2,000	$775,000	387.50	$150,000	0.19	9/27/00	5
Totals		**$126,000**	**$3,847,000**		**$1,445,000**			
Recovery or Salvage		**$66,000**	**$462,000**		**$295,000**	**0.64**		
Unrecovered/Unsalvaged Loss			**$3,385,000**					
Total Recovery/Salvage Cost + Unrecovered Loss			**$3,680,000**					

Cumulative Impact Loss vs. Recovery Cost Analysis

1	Lost Fabrication Drawings for Air Handlers	$175,000	$30,000	0.17	min leverage
2	Communication System Malfunction	$190,000	$70,000	0.37	
3	Leak in Tank Vessel	$237,000	$85,000	0.36	
4	Steel Support Misfabrication	$437,000	$280,000	0.64	
5	Switchgear Malfunctions	$462,000	$295,000	0.64	
6	Delayed Control System Approvals	$912,000	$870,000	0.95	
7	Unsubmitted Change Orders	$1,412,000	$920,000	0.65	
8	Electrical Contract Scope Errors	$2,197,000	$995,000	0.45	
9	Rigging Equipment Breakdowns	$3,072,000	$1,295,000	0.42	
10	Incomplete Survey of Existing Conditions	$3,847,000	$1,445,000	0.38	

Figure 8. Triage Matrix with Problems Listed in Order of Appearance

	Salvage Phase—Point of No Return								
I	J	K	L	M=J+L	N=M/D	O=M/F	P	Q	R=P–I
Days Remaining to Recover	Salvage Cost	Salvageability	Final Projected Loss	Loss Including Salvage Cost	Salvage Leverage	Salvage/ Recovery Cost Premium	Salvage Deadline	Days Remaining to Salvage	Breakeven/No Return Lag
8	$350,000	60%	$350,000	$700,000	0.80	2.33	11/7/00	46	38
2	$150,000	35%	$510,250	$660,250	0.84	8.80	12/5/00	74	72
5	$500,000	10%	$697,500	$1,197,500	1.55	7.98	9/30/00	8	3
10	$35,000	10%	$450,000	$485,000	0.97	9.70	10/25/00	33	23
1	$875,000	30%	$315,000	$1,190,000	2.64	2.07	10/15/00	23	22
3	$225,000	60%	$80,000	$305,000	1.53	1.56	12/2/00	71	68
9	$50,000	20%	$140,000	$190,000	1.09	6.33	10/15/00	23	14
13	$35,000	50%	$23,500	$58,500	1.24	3.90	11/20/00	59	46
2	$50,000	20%	$20,000	$70,000	2.80	4.67	10/12/00	20	18
6	$75,000	30%	$10,500	$85,500	5.70	2.14	9/30/00	8	2
	$2,345,000		**$2,596,750**	**$4,941,750**					
	$1,910,000		**$2,322,750**	**$4,232,750**	**1.25**	**3.68**			
				$709,000					

	point of diminishing returns

In this case study, the first five problems will be addressed; the last five are treated as a loss for this triage run. If their dates have not expired, then they can stay on the list and be included in the next triage run, along with the newly arrived problems. Alternately, they can be deferred to triage for salvageabilty or taken off the list.

The best recovery results in our case are achieved by triaging on the maximum loss leverage factor. For clarity, recovered problems 4, 7, 8, 9, and 10 are taken off the salvage triage list. In reality, an additional five problems (or ten or fifteen, depending on the particular project's constraints) would make the list for the next regular round of triage. All problems whose recovery deadline has not expired will be triaged for recovery, and at the same time, all problems whose deadline for salvage has not expired will be triaged for salvage. Thus, for example, problems 1, 2, 3, 5, and 6 may again compete for recovery resources with a new batch of problems, 11–15. If, however, we triage on a weekly basis, and our next run is on 29 September 2000, then the deadline for problems 5 and 6 will have expired. Knowing this time constraint, we may want to presort and address these problems first or decide to defer to the salvage phase.

Running Triage at the Point of No Return

As we know, the point of no return is the last chance to fully recover the project with the expenditure of maximum available effort. Therefore, at this point, the salvageability ranges from 0 percent to full recovery (100 percent), and the triage is performed both on the importance and salvageability criteria.

Now, let us look at the matrix criteria used for salvage triage (see Figure 13, page 62):

- Column J, Salvage Cost—This is the amount that must be spent to salvage the problem.
- Column K, Salvageability—This is an indicator of the percentage of the problem that can be salvaged.
- Column L, Final Projected Loss—If final projected loss is less than the Salvage Cost, and/or the Salvageability percentage is low, then we may want to write off this problem.
- Column M, final Loss, Including Salvage Cost—This cost only becomes a criterion if we decide to proceed with the salvage operations. If not, we write off the Final Projected Loss.
- Column N, Salvage Leverage—Equals to final Loss Including Salvage Cost (Column M) divided by Potential Impact Loss (Column D). It shows the bang for our salvage buck: the lower, the better.
- Column O, Salvage/Recovery Cost Premium—Equals to the Salvage Cost less the Recovery Cost. Even though Salvage Cost is often less than Recovery Cost, it is usually when the Salvageability is also low. This measure should be used prior to recovery triage, to determine the relative benefits of recovery versus salvage.
- Column P, Salvage Deadline date—This is the date of the point of no return. After this date, there is no possibility of a full recovery and the Salvageability decreases.
- Column Q, Days Remaining to Salvage.
- Column R, Breakeven/No Return Lag—This number indicates the float between the two points. Usually, the fewer days left, the more urgent it is to initiate recovery action rather than deferring for salvage.

Columns J through Q represent the criteria for running triage in the salvage phase. Triage is now run on both the importance and salvageability criteria. In the recovery stage, we set a constraint of being able to recover only five problems from the list of ten. Now, let us set a limit of two problems to salvage from the list of five. Before running the importance criteria sorts, it may be advisable first to run primary sorts for time and salvageability, and then to run secondary triage sorts on the importance criteria, as follows:

- Step 1 (presort if applicable): Column Q, minimum Days Remaining to Salvage.
- Step 2: Column K, maximum Salvageability—First, however, eliminate the upper (minor problems) and lower (terminal problems) percentage ranges, so that the range of Salvageability is, say, between 30 percent and 70 percent.

After the primary sort, run triage on the following criteria, with the objective of minimizing the total of salvage and loss costs:

- Step 3: Column M, minimum Loss Including Salvage Cost.
- Step 4: Column N, minimum Salvage Leverage.
- Step 5: Choose the least salvage and loss cost total for the two problems that you can salvage—i.e., triage on Step 3, which yields the lowest cost of $128,500. As you can see, the total of salvaged cost and unsalvaged loss is a constant, whether triaged on Step 3 or Step 4; in this case it is $594,000. However, note that the total final projected loss is only $509,000. Therefore, in this case, it seems best to write off all five remaining problems, since the total loss if we do nothing is less than the impact of any salvage effort.

Again, I must emphasize that this method is not a panacea for triaging all project management problems, but simply a template and an organized thought process. Each project is different, and the triage factors and resource constraints will vary. The most important thing to remember is to identify and address problems in a timely manner and practice the principles of triage. Also, estimating approximate impacts quickly is more important than precision. Fill in all the blanks based on the information that is readily available. Do not dwell on the gray areas or obsess over the details. Order-of-magnitude numbers and dates are OK. Correct relationship between various problem criteria is more important than absolute numbers.

This case study shows triage based on cost factors with the only constraint being the number of problems we can address in the recovery and salvage phases. More often than not, there are also budget constraints. Thus, if our recovery budget were only $200,000, we would have been able to address only the two top problems, 7 and 10, in our recovery triage (see Figure 10). Also, take a look at the comparative analysis of cumulative impact losses and recovery costs, as shown in the recovery triage examples. This analysis gives a cumulative recovery leverage factor in the order that the problems are to be addressed. Note that at the second problem, our cumulative recovery leverage factor is the lowest—i.e., our recovery bang for the buck is the greatest. Then, between problems 4 and 6, the difference in the cumulative factor makes the biggest jump upward, between 0.25 and 0.38. This indicates the biggest drop in the effect of recovery "bang:" we would have to spend an additional $575,000 to recover an additional $450,000 loss. Clearly, the payback from recovery efforts is gone at this point, even if we were not limited to the number of problems we can address.

Limits: Only Top Five Problems Can Be Addressed

Recovery Phase—Breakeven Point

Current Date 9/22/00

A	B	C	D	E=D/C	F	G=F/D	H
Problem	Description	Problem Potential Loss	Potential Impact Loss	Loss Leverage Factor	Recovery Cost	Recovery Leverage	Recovery Deadline
3	Leak in Tank Vessel	$7,000	$47,000	6.71	$15,000	0.32	10/5/00
5	Switchgear Malfunctions	$25,000	$25,000	1.00	$15,000	0.60	9/24/00
1	Lost Fabrication Drawings for Air Handlers	$25,000	$175,000	7.00	$30,000	0.17	10/1/00
2	Communication System Malfunction	$5,000	$15,000	3.00	$40,000	2.67	9/28/00
7	Unsubmitted Change Orders	$2,000	$500,000	250.00	$50,000	0.10	10/2/00
8	Electrical Contract Scope Errors	$15,000	$785,000	52.33	$75,000	0.10	9/24/00
10	Incomplete Survey of Existing Conditions	$2,000	$775,000	387.50	$150,000	0.19	9/27/00
4	Steel Support Misfabrication	$4,000	$200,000	50.00	$195,000	0.98	9/25/00
9	Rigging Equipment Breakdowns	$11,000	$875,000	79.55	$300,000	0.34	9/30/00
6	Delayed Control System Approvals	$30,000	$450,000	15.00	$575,000	1.28	9/23/00
Totals		**$126,000**	**$3,847,000**		**$1,445,000**		
Recovery or Salvage		**$64,000**	**$762,000**		**$150,000**	**0.20**	
Unrecovered/Unsalvaged Loss			**$3,085,000**				
Total Recovery/Salvage Cost + Unrecovered Loss			**$3,235,000**				
Cumulative Impact Loss vs. Recovery Cost Analysis							
3	Leak in Tank Vessel		$47,000		$15,000	0.32	
5	Switchgear Malfunctions		$72,000		$30,000	0.42	
1	Lost Fabrication Drawings for Air Handlers		$247,000		$60,000	0.24	
2	Communication System Malfunction		$262,000		$100,000	0.38	
7	Unsubmitted Change Orders		$762,000		$150,000	0.20	
0	Electrical Contract Scope Errors		$1,547,000		$225,000	0.15	min leverage
10	Incomplete Survey of Existing Conditions		$2,322,000		$375,000	0.16	
4	Steel Support Misfabrication		$2,522,000		$570,000	0.23	
9	Rigging Equipment Breakdowns		$3,397,000		$870,000	0.26	
6	Delayed Control System Approvals		$3,847,000		$1,445,000	0.38	

Figure 11. Triage on Minimum Recovery Cost

	Salvage Phase—Point of No Return								
	J	K	L	M=J+L	N=M/D	O=M/F	P	Q	R=P−I
maining over	Salvage Cost	Salvageability	Final Projected Loss	Loss Including Salvage Cost	Salvage Leverage	Salvage/ Recovery Cost Premium	Salvage Deadline	Days Remaining to Salvage	Breakeven/No Return Lag
	$500,000	10%	$697,500	$1,197,500	1.55	7.98	9/30/00	8	3
	$35,000	10%	$450,000	$485,000	0.97	9.70	10/25/00	33	23
	$350,000	60%	$350,000	$700,000	0.80	2.33	11/7/00	46	38
	$150,000	35%	$510,250	$660,250	0.84	8.80	12/5/00	74	72
	$225,000	60%	$80,000	$305,000	1.53	1.56	12/2/00	71	68
	$875,000	30%	$315,000	$1,190,000	2.64	2.07	10/15/00	23	22
	$50,000	20%	$140,000	$190,000	1.09	6.33	10/15/00	23	14
	$35,000	50%	$23,500	$58,500	1.24	3.90	11/20/00	59	46
	$75,000	30%	$10,500	$85,500	5.70	2.14	9/30/00	8	2
	$50,000	20%	$20,000	$70,000	2.80	4.67	10/12/00	20	18
	$2,345,000		**$2,596,750**	**$4,941,750**					
	$1,260,000		**$2,087,750**	**$3,347,750**	**1.07**	**4.35**			
				$1,594,000					

point of diminishing returns

Limits: Only Top Five Problems Can Be Addressed

Recovery Phase—Breakeven Point

Current Date 9/22/00

A	B	C	D	E=D/C	F	G=F/D	H	I
Problem	Description	Problem Potential Loss	Potential Impact Loss	Loss Leverage Factor	Recovery Cost	Recovery Leverage	Recovery Deadline	Days Re… to Rec…
10	Incomplete Survey of Existing Conditions	$2,000	$775,000	387.50	$150,000	0.19	9/27/00	5
7	Unsubmitted Change Orders	$2,000	$500,000	250.00	$50,000	0.10	10/2/00	10
9	Rigging Equipment Breakdowns	$11,000	$875,000	79.55	$300,000	0.34	9/30/00	8
8	Electrical Contract Scope Errors	$15,000	$785,000	52.33	$75,000	0.10	9/24/00	2
4	Steel Support Misfabrication	$4,000	$200,000	50.00	$195,000	0.98	9/25/00	3
6	Delayed Control System Approvals	$30,000	$450,000	15.00	$575,000	1.28	9/23/00	1
1	Lost Fabrication Drawings For Air Handlers	$25,000	$175,000	7.00	$30,000	0.17	10/1/00	9
3	Leak in Tank Vessel	$7,000	$47,000	6.71	$15,000	0.32	10/5/00	13
2	Communication System Malfunction	$5,000	$15,000	3.00	$40,000	2.67	9/28/00	6
5	Switchgear Malfunctions	$25,000	$25,000	1.00	$15,000	0.60	9/24/00	2
Totals		**$126,000**	**$3,847,000**		**$1,445,000**			
Recovery or Salvage		**$34,000**	**$3,135,000**		**$770,000**	**0.25**		
Unrecovered/Unsalvaged Loss			**$712,000**					
Total Recovery/Salvage Cost + Unrecovered Loss			**$1,482,000**					

Cumulative Impact Loss vs. Recovery Cost Analysis

10	Incomplete Survey of Existing Conditions	$775,000	$150,000	0.19	
7	Unsubmitted Change Orders	$1,275,000	$200,000	0.16	min leverage
9	Rigging Equipment Breakdowns	$2,150,000	$500,000	0.23	
8	Electrical Contract Scope Errors	$2,935,000	$575,000	0.20	
4	Steel Support Misfabrication	$3,135,000	$770,000	0.25	
6	Delayed Control System Approvals	$3,585,000	$1,345,000	0.38	
1	Lost Fabrication Drawings For Air Handlers	$3,760,000	$1,375,000	0.37	
3	Leak in Tank Vessel	$3,807,000	$1,390,000	0.37	
2	Communication System Malfunction	$3,822,000	$1,430,000	0.37	
5	Switchgear Malfunctions	$3,847,000	$1,445,000	0.38	

Figure 10. Triage on Maximum Loss Leverage Factor

	Salvage Phase—Point of No Return								
I	J	K	L	M=J+L	N=M/D	O=M/F	P	Q	R=P–I
Days Remaining to Recover	Salvage Cost	Salvageability	Final Projected Loss	Loss Including Salvage Cost	Salvage Leverage	Salvage/ Recovery Cost Premium	Salvage Deadline	Days Remaining to Salvage	Breakeven/No Return Lag
13	$35,000	50%	$23,500	$58,500	1.24	3.90	11/20/00	59	46
2	$50,000	20%	$20,000	$70,000	2.80	4.67	10/12/00	20	18
9	$50,000	20%	$140,000	$190,000	1.09	6.33	10/15/00	23	14
6	$75,000	30%	$10,500	$85,500	5.70	2.14	9/30/00	8	2
10	$35,000	10%	$450,000	$485,000	0.97	9.70	10/25/00	33	23
2	$150,000	35%	$510,250	$660,250	0.84	8.80	12/5/00	74	72
5	$500,000	10%	$697,500	$1,197,500	1.55	7.98	9/30/00	8	3
3	$225,000	60%	$80,000	$305,000	1.53	1.56	12/2/00	71	68
8	$350,000	60%	$350,000	$700,000	0.80	2.33	11/7/00	46	38
1	$875,000	30%	$315,000	$1,190,000	2.64	2.07	10/15/00	23	22
	$2,345,000		**$2,596,750**	**$4,941,750**					
	$245,000		**$644,000**	**$889,000**	**1.17**	**5.93**			
				$4,052,750					

point of diminishing returns

The triage matrix can also be modified to run triage based on schedule or on a combination of cost and schedule factors. In the case of schedule triaging, the potential impact loss would be limited to the maximum potential problem loss, indicating that the activity is on the critical path. The schedule triage criteria would be based on the principles of time/cost tradeoff, or expending the least amount of money to recover the maximum number of days. Details of this method are beyond the scope of this book.

Finally, for every triage operation, the question of unrecovered losses always remains: what do we do with the losses caused by problems that we can't recover or salvage? In the beginning of this case, I stated that the project in question was completed with an 8 percent profit. How can that be? With prompt identification of problems and aggressive triage, the financial impact of written-off losses can be minimized and the overall financial performance of the project improved.

Also, simply selecting a problem for recovery or salvage does not guarantee that the problem will in fact be recovered or salvaged. It simply means that we have used rational and logical methods to devote recovery and salvage resources to those problems that have the best chance of survival and the best survival value. Whether they actually do survive is a matter of proper implementation of emergency and crisis management techniques.

For those problems beyond our help, we must come to terms with the fact that they will result in a loss to the project. Triage is often the best method for project scope streamlining and selective elimination of "pet projects." Failing triage is often the best indicator that the activity should be canceled or terminated. However, this is not always possible; if you are a contractor on a lump-sum, fixed-scope project, you cannot arbitrarily eliminate a portion of that project simply because you are losing money on it. If scope performance is mandatory, then alternative cost-saving measures must be explored. These are commonly known as *value engineering*, and deal with achieving the same performance at less cost, details of which are beyond the scope of this book. Similar cost-cutting and performance-improvement methods are also discussed in various business process reengineering books.

In conclusion, Figure 15 (see page 63) features a flow chart that summarizes the entire project triage process.

Emergency Response

As defined earlier, if we are to draw a distinction between an emergency and a crisis, an emergency situation requires prompt reaction in order to prevent immediate and catastrophic consequences. An emergency response addresses immediate symptoms only; causes can be addressed later during the crisis-management phase, after the project is out of immediate danger.

Emergency Management Sequence

Emergency management involves five basic steps.

1. **Assess Recoverability**: We have previously discussed this in detail. You must know whether you are dealing with a recovery or a salvage operation, and what specifically you are trying to achieve with your emergency procedure.
2. **Conduct Triage**: Again, you must selectively focus your limited resources on those problem issues that are the most salvageable and have the highest value. Loss of the high-value functions can be critical to the overall system.
3. **Address the Symptoms—Apply Emergency Measures**: After triage, you must promptly and decisively apply emergency measures. At this point, it doesn't matter what caused the problem; you have to fix the problem now and stop the hemorrhage. Don't get sidetracked into intellectualizing the facts or analyzing why the problem has occurred.
4. **Seal Off the Damage**: After first aid has been applied, you must isolate the problem and its ability to spread and influence the entire system. In essence, this is a business quarantine.
5. **Maintain Vital Functions**: Overall operations of the system must continue during the "repairs." Most often, these operations can only be maintained at a minimum level until you "land the plane." In the meantime, you must compensate for the lost functions.

Emergency Management Techniques

Fixing immediate problems and maintaining vital functions is a multidisciplinary approach that blends analytical with intuitive thinking and relies heavily on experience and shortcuts, or heuristics. It can also be described as a scientific and systematic trial-and-error approach. You have a general idea of the actions that will stop the emergency, and you apply them in a sequence until the problem is resolved. Emergencies are typically "all-hands-on-deck" situations; the effort continues relentlessly until the situation is either stabilized or lost. Use all available tools, resources, and processes. Improvise; make do with what you have.

If something fails, the two questions that must be answered immediately are:

✔ What critical functions have failed?

✔ What can best replicate the failed function?

There are three basic techniques to replicate the failed function and keep the system running, as shown in Figure 16 (see page 64). These are the tools in your emergency toolkit—the baling wire, Band-Aid, and spare tire of business crisis management.

➔ Bypass provides a bridge around the failed part of the system, just like moving a crashed car to the shoulder so that traffic can pass. It is analogous to bypassing a failed blood vessel or splicing a wire to bridge a failed electrical connection.

➔ Patch involves fixing the failed part in place, assuming that the part is repairable and that the system can function during the patching process.

➔ Replace involves discarding the failed part and replacing it with a new part—assuming, of course, that the system can function during the replacement procedure and that the replacement part is available.

Salvage Phase—Point of No Return

I	J	K	L	M=J+L	N=M/D	O=M/F	P	Q	R=P–I
ys Remaining to Recover	Salvage Cost	Salvageability	Final Projected Loss	Loss Including Salvage Cost	Salvage Leverage	Salvage/ Recovery Cost Premium	Salvage Deadline	Days Remaining to Salvage	Breakeven/No Return Lag
13	$35,000	50%	$23,500	$58,500	1.24	3.90	11/20/00	59	46
2	$50,000	20%	$20,000	$70,000	2.80	4.67	10/12/00	20	18
6	$75,000	30%	$10,500	$85,500	5.70	2.14	9/30/00	8	2
9	$50,000	20%	$140,000	$190,000	1.09	6.33	10/15/00	23	14
1	$875,000	30%	$315,000	$1,190,000	2.64	2.07	10/15/00	23	22
	$1,085,000		**$509,000**	**$1,594,000**					
			$43,500	**$128,500**					
			$465,500						
				$594,000					

Salvage Phase—Point of No Return

I	J	K	L	M=J+L	N=M/D	O=M/F	P	Q	R=P–I
ys Remaining to Recover	Salvage Cost	Salvageability	Final Projected Loss	Loss Including Salvage Cost	Salvage Leverage	Salvage/ Recovery Cost Premium	Salvage Deadline	Days Remaining to Salvage	Breakeven/No Return Lag
9	$50,000	20%	$140,000	$190,000	1.09	6.33	10/15/00	23	14
13	$35,000	50%	$23,500	$58,500	1.24	3.90	11/20/00	59	46
1	$875,000	30%	$315,000	$1,190,000	2.64	2.07	10/15/00	23	22
2	$50,000	20%	$20,000	$70,000	2.80	4.67	10/12/00	20	18
6	$75,000	30%	$10,500	$85,500	5.70	2.14	9/30/00	8	2
	$1,085,000		**$509,000**	**$1,594,000**					
			$163,500	**$248,500**					
			$345,500						
				$594,000					

Limits: Only Top Five Problems Can Be Addressed

Recovery Phase—Breakeven Point

Current Date 9/22/00

A	B	C	D	E=D/C	F	G=F/D	H	
Problem	Description	Problem Potential Loss	Potential Impact Loss	Loss Leverage Factor	Recovery Cost	Recovery Leverage	Recovery Deadline	Days R… to Re…
8	Electrical Contract Scope Errors	$15,000	$785,000	52.33	$75,000	0.10	9/24/00	[illegible]
7	Unsubmitted Change Orders	$2,000	$500,000	250.00	$50,000	0.10	10/2/00	[illegible]
1	Lost Fabrication Drawings for Air Handlers	$25,000	$175,000	7.00	$30,000	0.17	10/1/00	[illegible]
10	Incomplete Survey of Existing Conditions	$2,000	$775,000	387.50	$150,000	0.19	9/27/00	[illegible]
3	Leak in Tank Vessel	$7,000	$47,000	6.71	$15,000	0.32	10/5/00	[illegible]
9	Rigging Equipment Breakdowns	$11,000	$875,000	79.55	$300,000	0.34	9/30/00	[illegible]
5	Switchgear Malfunctions	$25,000	$25,000	1.00	$15,000	0.60	9/24/00	[illegible]
4	Steel Support Misfabrication	$4,000	$200,000	50.00	$195,000	0.98	9/25/00	[illegible]
6	Delayed Control System Approvals	$30,000	$450,000	15.00	$575,000	1.28	9/23/00	[illegible]
2	Communication System Malfunction	$5,000	$15,000	3.00	$40,000	2.67	9/28/00	[illegible]
Totals		**$126,000**	**$3,847,000**		**$1,445,000**			
Recovery or Salvage		**$51,000**	**$2,282,000**		**$320,000**	**0.14**		
Unrecovered/Unsalvaged Loss			**$1,565,000**					
Total Recovery/Salvage Cost + Unrecovered Loss			**$1,885,000**					

Cumulative Impact Loss vs. Recovery Cost Analysis					
8	Electrical Contract Scope Errors	$785,000	$75,000	0.10	min leverage
7	Unsubmitted Change Orders	$1,285,000	$125,000	0.10	
1	Lost Fabrication Drawings for Air Handlers	$1,460,000	$155,000	0.11	
10	Incomplete Survey of Existing Conditions	$2,235,000	$305,000	0.14	
3	Leak in Tank Vessel	$2,282,000	$320,000	0.14	
9	Rigging Equipment Breakdowns	$3,157,000	$620,000	0.20	
5	Switchgear Malfunctions	$3,182,000	$635,000	0.20	
4	Steel Support Misfabrication	$3,382,000	$830,000	0.25	
6	Delayed Control System Approvals	$3,832,000	$1,405,000	0.37	
2	Communication System Malfunction	$3,847,000	$1,445,000	0.38	

Figure 12. Triage on Minimum Recovery Leverage Factor

Salvage Phase—Point of No Return

I	J	K	L	M=J+L	N=M/D	O=M/F	P	Q	R=P–I
emaining cover	Salvage Cost	Salvageability	Final Projected Loss	Loss Including Salvage Cost	Salvage Leverage	Salvage/ Recovery Cost Premium	Salvage Deadline	Days Remaining to Salvage	Breakeven/No Return Lag
	$150,000	35%	$510,250	$660,250	0.84	8.80	12/5/00	74	72
)	$35,000	10%	$450,000	$485,000	0.97	9.70	10/25/00	33	23
	$50,000	20%	$140,000	$190,000	1.09	6.33	10/15/00	23	14
	$500,000	10%	$697,500	$1,197,500	1.55	7.98	9/30/00	8	3
3	$35,000	50%	$23,500	$58,500	1.24	3.90	11/20/00	59	46
	$350,000	60%	$350,000	$700,000	0.80	2.33	11/7/00	46	38
	$50,000	20%	$20,000	$70,000	2.80	4.67	10/12/00	20	18
	$225,000	60%	$80,000	$305,000	1.53	1.56	12/2/00	71	68
	$875,000	30%	$315,000	$1,190,000	2.64	2.07	10/15/00	23	22
	$75,000	30%	$10,500	$85,500	5.70	2.14	9/30/00	8	2
	$2,345,000		**$2,596,750**	**$4,941,750**					
	$770,000		**$1,821,250**	**$2,591,250**	**1.14**	**8.10**			
				$2,350,500					

point of diminishing returns

Limits: Only Top Two Problems Can Be Addressed

Recovery Phase—Breakeven Point

Current Date 9/22/00

A Problem	B Description	C Problem Potential Loss	D Potential Impact Loss	E=D/C Loss Leverage Factor	F Recovery Cost	G=F/D Recovery Leverage	H Recovery Deadline
3	Leak in Tank Vessel	$7,000	$47,000	6.71	$15,000	0.32	10/5/00
5	Switchgear Malfunctions	$25,000	$25,000	1.00	$15,000	0.60	9/24/00
2	Communication System Malfunction	$5,000	$15,000	3.00	$40,000	2.67	9/28/00
1	Lost Fabrication Drawings for Air Handlers	$25,000	$175,000	7.00	$30,000	0.17	10/1/00
6	Delayed Control System Approvals	$30,000	$450,000	15.00	$575,000	1.28	9/23/00
Totals		$92,000	$712,000				
Recovery or Salvage		**$92,000**	**$712,000**				
Unrecovered/Unsalvaged Loss							
Total Recovery/Salvage Cost + Unrecovered Loss							

Figure 13. Triage on Minimum Loss and Salvage Cost

Limits: Only Top Two Problems Can Be Addressed

Recovery Phase—Breakeven Point

Current Date 9/22/00

A Problem	B Description	C Problem Potential Loss	D Potential Impact Loss	E=D/C Loss Leverage Factor	F Recovery Cost	G=F/D Recovery Leverage	H Recovery Deadline
1	Lost Fabrication Drawings for Air Handlers	$25,000	$175,000	7.00	$30,000	0.17	10/1/00
3	Leak in Tank Vessel	$7,000	$47,000	6.71	$15,000	0.32	10/5/00
6	Delayed Control System Approvals	$30,000	$450,000	15.00	$575,000	1.28	9/23/00
5	Switchgear Malfunctions	$25,000	$25,000	1.00	$15,000	0.60	9/24/00
2	Communication System Malfunction	$5,000	$15,000	3.00	$40,000	2.67	9/28/00
Totals		$92,000	$712,000				
Recovery or Salvage		**$92,000**	**$712,000**				
Unrecovered/Unsalvaged Loss							
Total Recovery/Salvage Cost + Unrecovered Loss							

Figure 14. Triage on Minimum Salvage Leverage

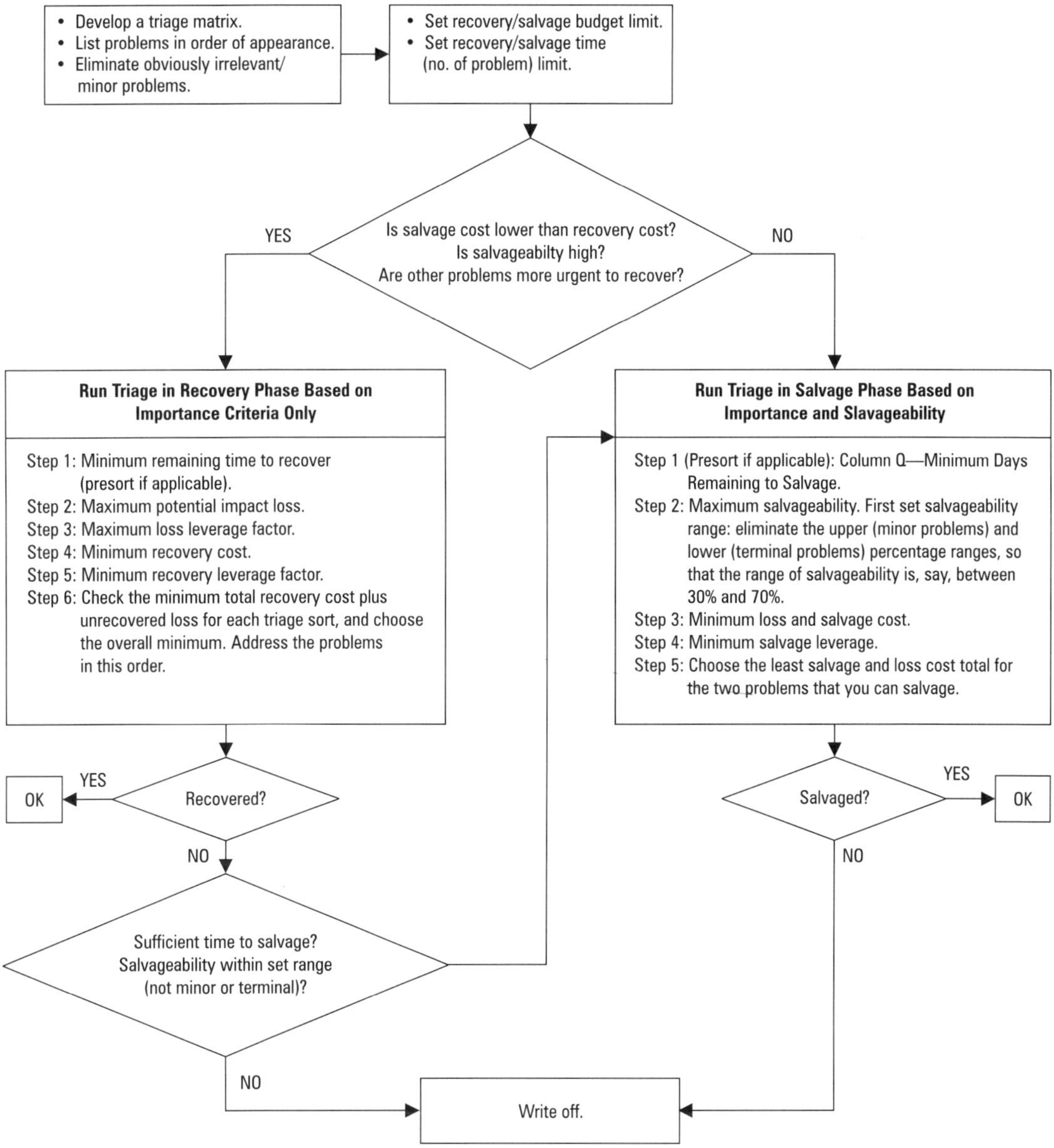

Figure 15. Project Triage Flow Chart

Most often, the emergency fix procedure involves a combination of all three techniques, due to the urgency of the situation and the limited effectiveness of each technique applied alone. There are two sequences in which the emergency fixes can be applied: serial and parallel, as illustrated in Figures 17 and 18 (see pages 64–65).

The serial model involves applying techniques one after the other. If the first technique fails, or is not completely successful, then the second technique may be a "stronger dosage" of the first, or a totally different approach.

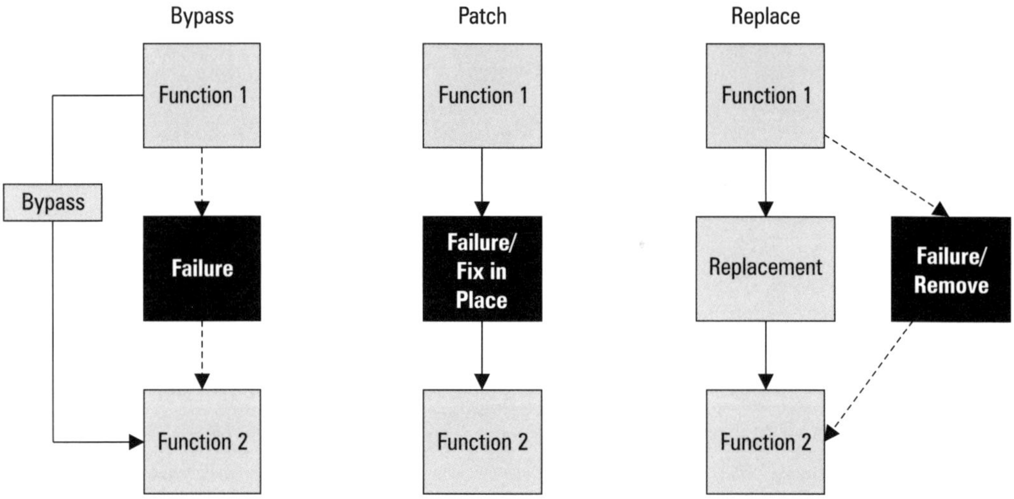

Figure 16. Emergency Management Techniques

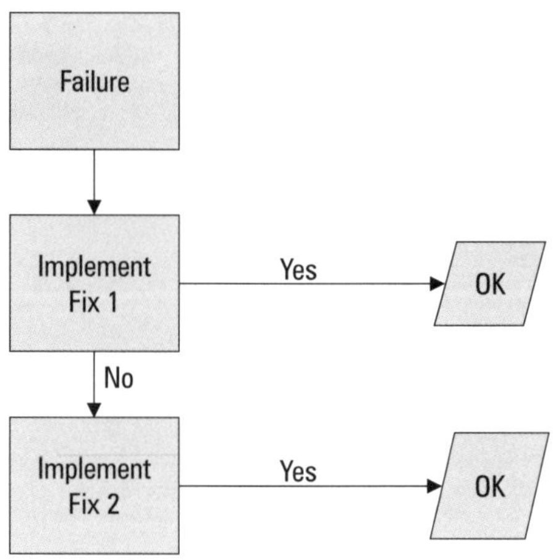

Figure 17. Emergency Fix—Serial Sequence

Sequence of application is especially important in the serial model. The fixes must be applied in a correct sequence; otherwise, there could be more damage than good. If immediate cure cannot be achieved on the first try, then the objective is incremental improvement with each successive fix application. The "if it doesn't kill it, it will cure it" approach should be avoided.

In a parallel model, two or more fix techniques are applied at the same time. The complementary effect of the fixes applied together is the primary concern. All concurrently applied techniques must be mutually compatible—there can be no "counter indications."

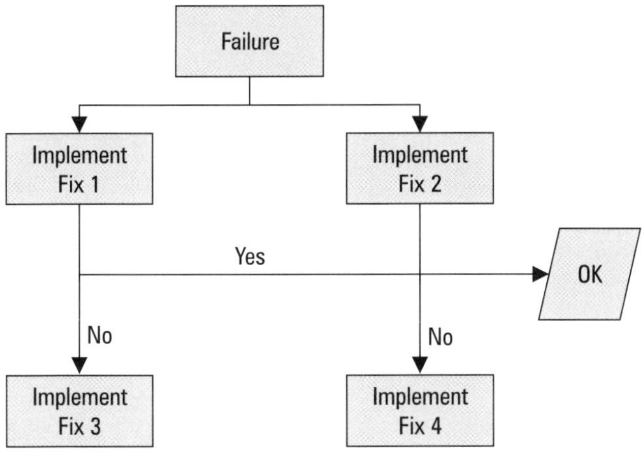

Figure 18. Emergency Fix—Parallel Sequence

This brings us to the conclusion of the emergency management portion of our effort. We have learned how to quickly assess a crisis situation, conduct triage, establish recovery or salvage objectives, and perform emergency repair procedures. We hope that the hemorrhage has been stopped and the business problem is out of the emergency room and in intensive care, able to perform vital functions while connected to the financial and management life-support system. Our focus can now shift from fixing the symptoms to curing the causes of the crisis.

PART II

Crisis Management

Intensive Care—Addressing the Causes

Causes of Crisis

Once the company or project is out of the emergency room and into intensive care, the focus shifts from addressing the symptoms to addressing the causes of the crisis. To reiterate a previously made point, main causes of business crises are human—there are very few real acts of God or *force majeures* in business. Most "technical" causes can also be ultimately traced to human error.

What we see as the obvious manifestation of problems are the symptoms, not the causes. Let us list some common crisis symptoms:

- High employee turnover.
- Cash-flow problems—inability to meet current financial obligations.
- Schedule delays.
- Budget overruns.
- Low productivity/low revenue per employee.
- Accidents.

However, in order to address these symptoms, we must solve their underlying causes. Some major causes of a crisis can be classified as follows:

- Failure to recognize and address errors.
- Failure to make decisions.
- Failure to learn from mistakes (over and over again).
- Incompetence/lack of expertise/lack of problem-solving skills.
- Lack of accountability—the buck doesn't stop anywhere.
- Lack of attention to details.
- Procrastination; allowing small problems to fester and grow.
- Failure to recognize the warning signs.
- Ignoring the warning signs and hoping that the problems will go away.
- Failure to plan, follow through, and expedite.
- Failure to start, sustain, and complete a task.

- Discontinuity of responsibility—"I was not involved."
- Organizational bottlenecks—systems and procedures/multiple layers of management.
- Unnecessary steps/inefficiency.
- And so on.

In human (or corporate) behavior terms, causes of a crisis are usually manifested as problem avoidance:

- Ignore it.
- Pass it to someone else.
- Avoid responsibility.
- Deny involvement.
- Drop the ball in the middle.
- Pass the blame/explain/exculpate.

Unfortunately, very few problems resolve on their own—there are no miracle cures. Abandoning problems makes them someone else's. Although in some situations a "do nothing/wait and see" attitude is acceptable, these cannot be classified as crises. In an emergency or a crisis, things get worse if no action is taken. This is no argument for rash judgment and action; remember: first do no harm, and make no errors in a hurry.

The main causes of all crisis situations can be boiled down to four critical resources:

Resource	Problem
Money	(Lack Thereof)
Time	(Late)
People	(Incompetent)
Information	(Inaccurate)

Perhaps the first two are essential, because, given enough time and money, good people and information can usually be obtained. All other resources/problems specific to a particular industry can usually be resolved, given enough time and money with good people and accurate information: production breakdowns, logistics bottlenecks, etc.

Catastrophe Theory Applied

All systems—whether natural, mechanical, or organizational—search for stability, equilibrium, or what is also called the state of lowest total energy. Catastrophes occur when a system loses stability: something tips it over the edge, it loses balance, and then it falls abruptly until it finds a new stable point. Think of the examples in nature, such as volcanoes and earthquakes.

In a stable system, a change in input is proportional to the change in output—they move in sync according to some established formula, if not linearly. We come to expect or "feel" normal responses from things and events in everyday life. We add a little more pressure to the gas pedal, and the car accelerates a little. We put another book on the shelf, and the shelf deflects a little more. Catastrophes or failures occur when something whacks this "normal" system out of balance, when the equilibrium is disrupted. A small, incremental change of input then causes a disproportionate output or response, a collapse—the proverbial straw that broke the camel's back.

What Causes the Systems to Fail?

Material science and engineering provide an example. When material samples—say, steel rods—are subject to tension, they deform or stretch. Up to a certain limit of tension force, the deformation is elastic; in other words, the rod returns to its original shape. After additional force is applied, the rod stretches but does not return to the original shape—it becomes "plastic." Finally, with enough force, the rod snaps in two. Different materials have different elastic and plastic limits. A rubber band has a high elastic but low plastic limit.

To summarize:

➔ Elastic limit—returns to original state after force is applied.

➔ Plastic limit—does not return to the original state; yields and deforms ("stretches") under a force, but does not break.

Why is this concept important to business? Because the analogy widely applies—organizations have built-in elastic and plastic limits that allow them to roll with the punches, compensate for problems, and recover from stress. After the elastic limit is exceeded and we are in the plastic zone, we'll "never be the same again." What are the elastic and plastic limits for critical functions in your organization? Under what stresses can your company return to normal—bend but not break?

How a business system reacts to crisis stress depends in a large measure on the internal ability to compensate. People work harder, longer hours; cash reserves mask the financial hemorrhages. However, there are no warning signs, no mechanisms to alert us to the problem and initiate a corrective procedure. What finally happens is a sudden "dissonance resolution"—sudden because the system has "invisibly," without outward signs of a crisis, compensated for the stresses until a sudden rupture occurs. We all know there are problems and conflicts, but we compensate and "keep it all in" until the endurance limit is reached. The trick is to combine high organizational resiliency—the "elasticity" and "plasticity"—with early warning mechanisms that prevent a crisis from festering out of sight. We'll discuss this more in Part III—Crisis Prevention.

Failure Modes and Effects

Why do things fail or go wrong? Given time, every component in a system will eventually fail, whether it is a business, a computer, or a human body. Business systems can be maintained for long periods of time if we can predict and prevent or minimize the impact of component failures. This subject will be discussed in detail in the Crisis Prevention section of this book, but here are some basic outlines as they pertain to crisis diagnostics and symptom recognition.

To understand where a crisis will probably strike, and what effect it will have, we have to understand how the entire system operates. This is a two-step process.

Develop a System Functions Map

Any process—small or large, simple or complex—can be graphically depicted as a network of interconnected functions, with functions shown as nodes and relationships as lines in a network. Think of the function or activity nodes as "tanks" and of connecting lines as plumbing, with work flowing through the system.

All "work" in a business system can be defined as productive, contributory, or nonproductive. Productive work is value added; in other words, it creates something. Contributory work does not directly generate any revenue, but it provides support for contributory work. Nonproductive work is just that—it adds no value and often eats the revenue generated by the productive work. (Often, nonproductive, needless activities are disguised as "contributory work"—e.g., supervision, review, etc.) Here are some examples:

Productive	Contributory	Nonproductive
Make	Review	Procrastinate
Create	Assist	Criticize
Invent	Comment	Delay
Work	Support	Obstruct

Numerous symbols and methods can be used to graphically map a system or a work process. Think of computer networks, air-route maps, football diagrams, etc. The simplest form, using only four symbols, is commonly used with minor variations in the business-process mapping for total quality management (TQM) and reengineering. The four symbols, with their respective value designations, are (see Figure 19):

- Activity—productive
- Transport—contributory
- Inspection—contributory
- Delay—nonproductive

Activity or processor is any productive work performed toward accomplishing the business objectives. It could be cutting steel sheets, laying bricks, making a decision, or any other value-added activity that directly contributes.

Transport is physical movement of the product: things, ideas, power, whatever—whether by air, road, forklift, fax, pipeline, wire, or other means.

Delay or storage is unproductive work, sometimes unavoidable due to bottlenecks in the system, or slower rates of processing downstream. Inspection may be defined as necessary delay, ensuring that all products are correct before shipping. However, if production is on schedule but half of our trucks have broken down, we have a bottleneck delay.

The standard flow-chart symbols in Figure 19 can express any business process and can be time-scaled on an x-y graph, or space scaled on a map. Overall function of the system can be simulated with the use of computer software to predict how it will behave under different conditions and constraints. This process is also known as "system dynamics."

Define Failure Modes and Critical Points

Once we have a systems function map established, we can begin to study the different paths along which the crisis impact will "travel" through our system and which components it will affect.

The path that a crisis takes through a system is called a *failure mode*. All failure modes in a system represent a fault tree, a backtracking path from each potential failure to each potential cause. In TQM literature, informal

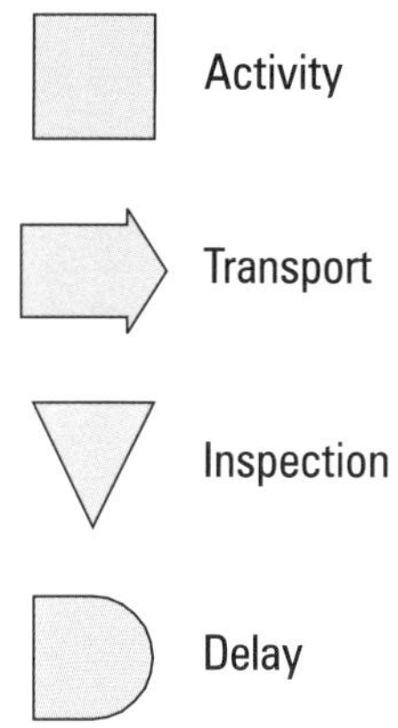

Figure 19. System Function Symbols

fault trees known as "fishbone" (due to their shape) or Ishikawa diagrams are used as informal brainstorming tools. The difference is that a "fishbone" diagram does not map out and follow the system path.

Analyzing a fault tree answers the following questions: What part of the system can fail? Is it hidden or predictable? If disaster strikes component A, and the impact flows down paths (failure modes) 1 and 2, how will components B and C be affected? It depends on a number of factors, but most importantly the following:

Strength of the Component:
Ability of affected component to handle the impact.

Importance of the Component:
How critical is the component to the entire system?
If it fails, will the entire system fail?

If the component in question is very important but not very strong, we have a problem. Such a component can cause catastrophic system failure, since it is a link in the system chain that is both critical and weak.

Components with *high importance* and *high failure effect* are called the *critical points* in the system and must be protected from the direct impacts of crises. To be critical, a component must satisfy the following two conditions:

➔ High Importance = Component performs vital functions.
➔ High Failure Effect = Component is indispensable and cannot be easily replaced.

In mechanical systems, a critical component could be a crucial relay that controls power supply to a number of important components. In business organizations, critical (weak?) links are usually people—and the best people at that. Look around your organization: best people tend to be overloaded and stressed while performing crucial tasks.

Conditions for a critical failure can be summarized as follows:

- Overloaded system—lowered processing or handling capacity.
- Low safety margin/margin of error.
- Low redundancy—no shortcuts or bypasses.

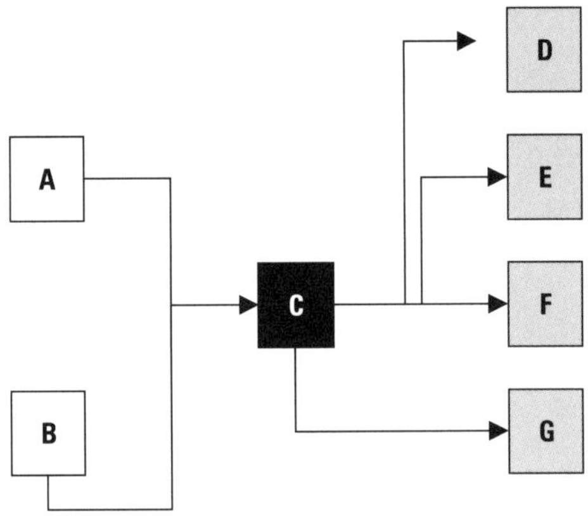

Figure 20. System Function Map

- Low backups—unique component, no stand-ins.

From a business systems standpoint, these are the primary symptoms manifestations:

→ Bottleneck—slow processing at a critical point; either flow too fast or processor too slow.
→ Information overload—unaddressed problems sit in a pile.

In manufacturing, resolving production bottlenecks and constraints is achieved by methods known as "line balancing," where the objective is to smooth the flow of materials through the production system. The project management environment is significantly more complex than a process environment with more variables and less predictability. Let us call this failure prevention method "system optimization." Here are its basic characteristics:

- Eliminates nonproductive activities.
- Reduces contributory activities to the absolute minimum needed to support the productive activities.
- Keeps the activity flow; avoid backtracking.
- Resolves delays and bottlenecks.
- Multiple channel processing.
- Increases speed of bottleneck processor.
- Replaces/"opens another line" or bypasses bottleneck processor.

Failure modes all converge through component C (see Figure 20); it is the critical component. Think of items A or B as power sources, item C as a switch, and downstream items D through G as lights. If A goes out, there is still B to provide power. If D goes out, there are still three more lights on. But if C fails, all lights are out, regardless of the fact that A and B are still generating power. C is a critical component at the convergence of the system.

Summary

The following diagnostic steps must be taken to identify the failure modes and critical components in your system:

✔ **Map critical system functions**.
✔ **Define failure modes**.
✔ **Determine effects of failures**.

✔ **Work on weak links**:
- Provide backups and redundancies.
- Isolate effects—decrease criticality.

✔ **Fix weakest links first, those**:
- With worst possible outcomes/effects.
- Affecting the greatest number of other downstream components.
- With a greatest number of failure modes—under most "stress."

Critical functions must be kept out of the failure mode paths. This is seldom possible in practice; highly important functions and people are, by definition, in the thick of the battle. The crucial requirement is to provide them with adequate resources and support so that they can withstand the stresses. Restructuring your business organization to divert crisis impact from the most important components is a priority.

Diagnostic Methods

As in medicine, so it is in business; an accurate diagnosis is essential in determining the cause of a problem. Problem cause must be diagnosed in order to develop an effective solution. The alternative is an ineffective trial-and-error approach where solutions are randomly thrown at the problem in hope that something will work, often causing more harm than good. You may recall, however, that the trial-and-error and rule-of-thumb methods are appropriate in emergency situations—we are applying a limited number of rudimentary procedures to address immediate symptoms in situations where speed takes precedence over perfection.

Diagnostic methods can be broadly classified as active or passive. Think of active methods as hunting and passive methods as trapping. When hunting, you actively pursue a specific prey. When trapping, you set a trap and then return to see what, if anything, you caught.

Needless to say, active methods are preferred in a crisis situation, although passive methods can be set as a part of a crisis-prevention program to provide an early warning about a problem. Most passive diagnostic methods are traps or "tests" set for a specific cause. For example, you may want to know when your monthly sales for a specific product decline below a certain level and set your business information system to warn you accordingly, but you are not actively pursuing this data. Passive methods are therefore great timesavers for noncrisis situations, allowing you to focus on more pressing matters, while the system only alerts you when something goes wrong. Crisis prevention as an organized approach will be discussed in detail in a later section.

In most crisis situations, we will be forced to actively analyze the symptoms and determine the causes with a reasonable degree of confidence. There are three basic active diagnostic methods:

1. **Failure mode analysis.**
2. **Elimination.**
3. **Cause isolation.**

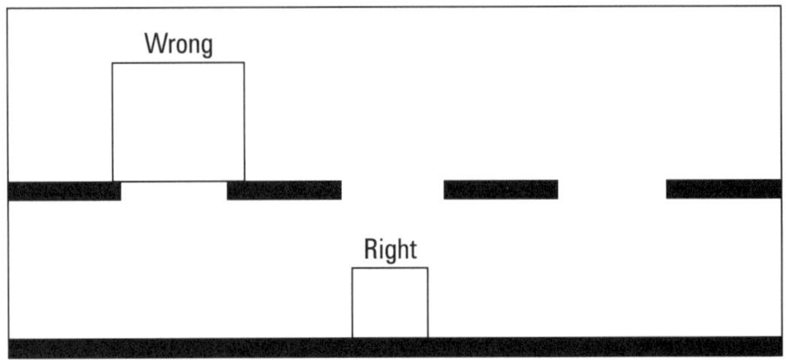

Figure 21. Diagnostic Test by Elimination

Failure Mode Analysis

This approach was discussed in detail in the Failure Modes and Effects discussion. It works well if a system functions map exists and a fault tree can be developed reasonably quickly. A variation of this method is to develop a fault tree based not on the entire system, but only on the specific decisions made preceding the problem. In essence, the fault tree now becomes a backward decision tree, and it allows you to backtrack your decisions, see where you took a wrong turn in the road, and isolate the fault.

Elimination

> I ran out of mistakes to make.
>
> Thomas Edison

As the name implies, diagnostics by elimination involves successively eliminating data that does not pass the test. The elimination test mechanism is set as a screen or a sieve.

There are two types of diagnostic tests by elimination:

- ➔ Type I—eliminate what is right; what remains is wrong.
- ➔ Type II—identify what is wrong; what remains is right.

Type II is shown schematically in Figure 21.

Note that this method eliminates, on a pass/fail basis, one criterion at a time. It is therefore important to set the elimination screens in a correct order, and it is usually necessary to repeat the process several times to isolate the cause. To paraphrase Sherlock Holmes: If we eliminate that which we are certain is not the cause, then what remains must be the cause.

Cause Isolation—The 80/20 Rule or Pareto Principle

The Pareto principle is named after its Italian inventor, who observed that 80 percent of land in medieval Italy was held by 20 percent of the population. Over time, others have discovered that this ratio holds for a wide range of social phenomena, if only anecdotally:

- Twenty percent of the employees seem to do 80 percent of the work; it follows logically that the remaining bulk of slackers only do 20 percent of the work—does this sound true in your company?
- Twenty percent of customers provide 80 percent of the business.

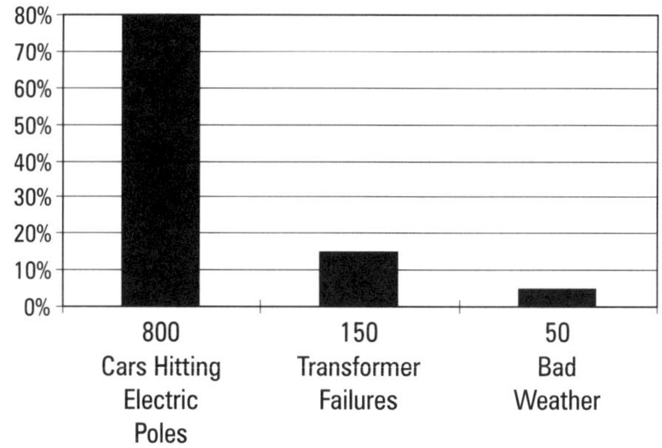

Figure 22. Cause Isolation—Step 1

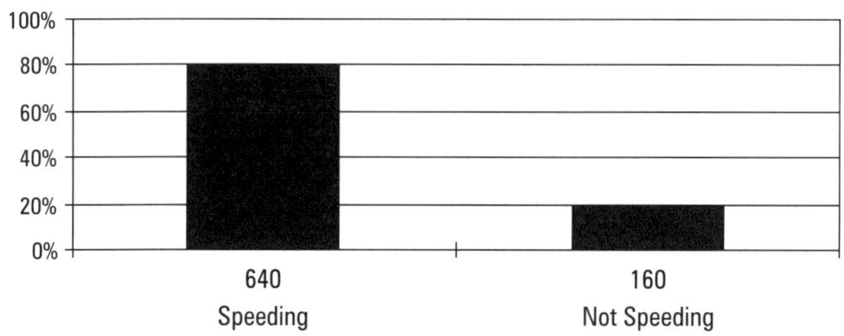

Figure 23. Cause Isolation—Step 2

- Twenty percent of the products generate 80 percent of the sales.
- And so on—you get the idea.

Now, why is the Pareto Principle important as a diagnostic tool? Because it allows us to sequentially eliminate irrelevant data and isolate the cause of a problem. Consider the following allegedly true example, simplified here for brevity.

You are a local manager of a power utility. You are faced with an unprecedented number of power failures along one distribution line—1,000 in the last year. Where do you start investigating the cause? First you collect the data to determine the major cause of the failures. You group the initial data and notice that 80 percent of the failures are caused by cars hitting the electric poles (see Figure 22). Now you know your focus, instead of immediately modernizing the transformer equipment or sending crews out on standby in thunderstorms.

But, where do accidents occur? Do cars randomly hit the poles? Were the drivers impaired? Was the weather bad? All these factors could play a part, but none account for more accidents than the others. You notice the following major cause: 80 percent of the drivers hitting your electric poles were speeding (see Figure 23).

But speeding alone doesn't make the car veer off the road, and you have already eliminated weather and drinking as significant contributing factors.

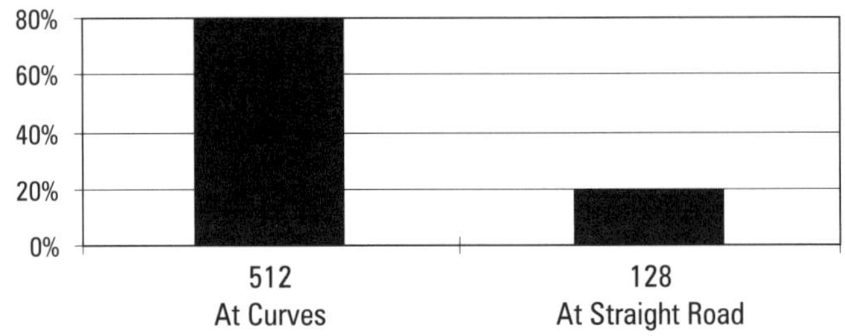

Figure 24. Cause Isolation—Step 3

You focus on the location: 80 percent of the poles hit were located at curves. The drivers lost control of their vehicles and veered off the road in a straight line into the poles (see Figure 24).

Things now make a lot more sense. Over 50 percent of all power outages were caused by speeding drivers who lost control of their cars in curves and hit the poles that happened to be located at a tangent to the road. Location of the poles is as much of a problem as speeding drivers. There is nothing you can do about the drivers, but you can relocate the poles away from the trajectory of veering cars. Thus, with focused analysis and minor adjustments, you have effectively isolated the main cause of your problem and reduced its impact by half. You can now focus on the second most critical cause, and so on. Contrast this to a disorganized trial-and-error approach where you randomly throw solutions at problems, hoping that the problem will disappear.

Data Collection and Analysis

> First get your facts straight, then you can twist them all you want.
>
> Mark Twain

You are now ready to start collecting the data that you will need to develop the solution to the crisis.

Data collection and analysis is ongoing; it is conducted at all stages of the crisis management process. Determining both the symptoms and the causes involves collection of data. However, after diagnosing the major causes, you can set a *problem statement*—a formula or an equation that defines the problem and identifies the missing information required to solve that problem. Your task can now focus on finding the missing information and converting raw data into *facts*. The objective is to reveal the missing truth.

Again, as in the search for causes, the effort of collecting data for solution development can be passive or active. In most crisis situations, you won't encounter a lack of data; to the contrary, you will likely be bombarded with streams of "data smog." The trick is to weed the useless information from the raw data material (80 percent?!) until all that remain are facts. This weeding process to get to the crux of an issue is the focus of the data analysis process.

The process of data collection, analysis, and synthesis is not unlike the legal cross-examination process. Criminal defense lawyer F. Lee Bailey identifies the following three phases of cross-examination:

- ➔ Extraction—extracting the facts from a witness.
- ➔ Closing—closing the avenues of escape, eliminating the possibility for the witness to weasel out.
- ➔ Impeachment—confronting the witness' contradictions and exposing the lies.

Data and Its Manifestations

Data in the context of business crisis management refers to all the information related to the crisis. Data can:

- Reflect the symptoms.
- Define the problem causes.
- Give us a clue to the solution.

Data pertaining to business problems comes in many formats—basic is alphanumeric; it is either expressed as numbers or words. Data can appear as a random, ambiguous manifestation from multiple sources, or we can actively search for the specific data that we need, assuming that we know what we are looking for and where to find it. A crisis manager must be comfortable with operating within ill-defined ambiguous problems.

Data manifestations are:

- Either obvious—bombarding you.
- Or hidden—you have to look.
- Most often a combination of both.

Before devising the right solution, we must ensure that we are tackling the right problem.

Identify and Define the Problem

What is a problem?

- A *problem* is the difference between reality and the desired state of affairs, or the difference between actual accomplishment and planned objective.

Set the Problem Statement

- ✔ Define the "as-is" current actual status/progress of events.
- ✔ Define the planned status of events—where should you be now?
- ✔ Compare the actual versus planned status—how bad is the discrepancy (refer to recoverability)?

For example, let us say that the required daily production is twenty-four units, or three units per hour. You have three machines that should be capable of producing this output, but you can only produce fifteen units.

Machine	Planned Production (M)	Breakdown Percentage	Operational (P) Percentage	Actual Production
1	10	40%	60%	6
2	8	25%	75%	6
3	6	50%	50%	3
Total	**24**			**15**

Why? You have determined that they all break down a certain, varying percentage of the time, and you can set your problem statement, i.e., the formula, as follows:

$$Output = (M1 \times P1) + (M2 \times P2) + (M3 \times P3)$$

Now you know that you must focus your analysis on the causes of breakdowns, not on some other issue that is not a factor—for example, labor shortage or lack of skills or raw materials. If these factors were influencing productivity, you would include them in the formula as variables. To reiterate, you only need the minimum number of facts required to develop a solution. If the problem statement included the plant manager's shoe size and the machine operator's weight, there are people who would try against all reason to find the relevance of this information to the solution.

Specify the Solution Requirements

- ✔ What is the solution method or action supposed to accomplish?
- ✔ Define the problem formula or equation—data input and output, constraints, problem variables and their relationships.
- ✔ Determine the minimum number of variables needed to solve the problem.
- ✔ Identify unknown and necessary variables—focus data search on these variables.

Convert Raw Data into Facts

- ✔ Collect the raw data/information.
- ✔ Eliminate information overload—filter the useless data, "background noise," and "data smog."
- ✔ Focus on important things only, the crux of the issue.
- ✔ Zero in on the root of the problem.
- ✔ Set "screens" and "traps" to weed facts from useless information.
- ✔ Reduce data to its essential significance.
- ✔ Convert data into the absolute minimum number of facts required to develop a solution.
- ✔ Base analysis on facts; minimize assumptions, guesses, and round numbers.

Data Analysis

The outlined process is not a mandatory sequence. You will be going back and forth between steps depending on the particular nature of the problem. However, the process provides a structured approach where each step builds on its predecessor to sequentially reduce the data to its minimal factual essence, the lowest common denominator for solution development.

Steps 1 and 2 eliminate the obviously erroneous or irrelevant data. Step 3 projects the outside influences that will change the values and formats of the data as we continue the analytical process. Steps 4 and 5 eliminate the verbal inconsistencies and errors, and finally, steps 6 and 7 weed out the numerical misrepresentations. To summarize:

Step 1. Eliminate irrelevant data.
Step 2. Prioritize relevant data in terms of importance.
Step 3. Adjust relevant data for factual drift.

Step 4. Eliminate logical fallacies.
Step 5. Eliminate doublespeak and doublethink.
Step 6. Eliminate false/irrelevant trends.
Step 7. Eliminate statistical lies.
Let us look at each step in detail.

Step 1—Eliminate Irrelevant Data

I keep six honest serving-men (they taught me all I knew); their names are What and Why and When and How and Where and Who.

Rudyard Kipling

We may notice logical fallacies right away, but if they are irrelevant, ignore them—don't waste time.

- ✔ Does it matter? If it matters, how much? Is it relevant or required to develop the solution?

The field of consciousness is very narrow.

St. Exupéry

Focus! The human mind can parallel-process no more than seven pieces of information; most people are "serial"—they deal with one problem at a time.

- ✔ You need only the minimum amount of data required to formulate the problem and develop a solution. Sift through the data smog.
- ✔ Prioritize relevant data in terms of relative importance.
- ✔ Ask *why* until the base question is answered—like a two-year-old would. Peel the layers of "because" responses until the root question is answered.

Step 2—Prioritize Relevant Data in Terms of Importance

Show your data in a chart or a matrix format for easy comparison.

Charts are the tools to explain the obvious to the ignorant.
Or, a picture is worth a thousand words?

- ✔ Compare/evaluate facts, data, options, costs, and benefits against the same set of criteria, on an equalized basis.
- ✔ Is your comparison direct—all factors have the same weight, or use a weighted scorecard?

Step 3—Adjust Relevant Data for Factual Drift

- ✔ Compensate for factual bias and changes in time, space, and data values.
- ✔ Check trends and estimate the value/location of data at the time needed. Aim toward where your target is going to be.

Step 4—Eliminate Logical Fallacies/Obstacles to Logical Thinking

When I use a word, it means exactly what I choose it to mean, neither more nor less.

Humpty Dumpty to Alice in Wonderland

Fallacies represent faulty reasoning and illogical thinking by the use of language (words and data) to misrepresent reality or deviate from truth, whether intentionally or not.

✔ Unintentional when used by the incompetent to justify their points of view.
✔ Intentional when used by professional manipulators/spin-doctors.
✔ Fallacies present major obstacles to logical thinking and to your pursuit of the truth, and must be eliminated.
✔ Ask yourself: Does it make sense? Always remember: reality is what it is, and not what we wish it to be.

Numerous categorizations of fallacies exist in the logic and philosophy theories, but they can all be grouped into four basic types and numerous subgroups:

1. **Fallacies of authority**.
2. **Fallacies of distraction**.
3. **Conceptual fallacies**:
   - Fallacies of ambiguity.
   - Missing the point.
4. **Reasoning fallacies**:
   - Inductive fallacies.
   - Fallacies involving statistical syllogisms.
   - Causal fallacies.
   - Category errors.
   - *Non sequitur* (does not follow).
   - Syllogistic errors.
   - Fallacies of explanation.
   - Fallacies of definition.

Abundant literature exists that deals with the subject matter. In this book, we are only going to focus on salient examples that will enable you to quickly identify and refute the major fallacies. Try by filling in the Description and Rebuttal boxes where blank in Table 3.

Step 5—Eliminate Doublespeak and Doublethink

Whereas logical fallacies represent the misuse of language designed to twist logical thought, doublespeak is an intentional misuse of language designed to misrepresent reality. A speaker guilty of using logical fallacies may be excused, due to incoherent thinking. A doublespeaker, on the other hand, is a demagogue. He has perfectly lucid thoughts and knows the truth about the subject matter, but chooses instead to mislead the listener, obfuscate reality, and deflect probing questions.

To summarize the difference between logical fallacies and doublespeak:

➔ Logical fallacy = faulty reasoning—not always intentional.
➔ Doublespeak = obfuscation of reality—almost always intentional.

Author William Lutz in his 1989 book, *Doublespeak,* (New York: Harper & Row) identifies four types of doublespeak:

- Euphemism or understatement.
- Jargon.
- Gobbledygook or bureaucratese.
- Inflated language.

Fallacy	Description	Argument	Rebuttal
Authority			
Appeal to Authority—Also Hearsay	Invoking unnamed *authority* to lend credence to an argument.	The experts agree that this is the preferred manufacturing procedure.	Who are these experts? What makes them experts? Where did you hear this?
Appeal to Consequences	Bad things will happen if you don't agree.	You can't say that this project will end in failure because it makes you a defeatist and a loser.	On the other hand, you may just be a realist.
Prejudicial Language		All loyal employees will agree that a severe workforce reduction is necessary to ensure the survival of the company.	Implication is that if you disagree, you are disloyal, which may not necessarily be true.
Appeal to Popularity		Everyone in the company knows that this is the correct data, so why are you trying to stand out and deny it?	
Distraction			
Style over Substance		We should trust the other guy's data. He is a much better dresser.	
Argument from Ignorance		Since the audit data cannot prove that the company will go bankrupt, it probably won't.	It may go bankrupt regardless.
Sanctioning the Devil	Ducking the issue by pretending it is beneath you to discuss it.	I'm not even going to argue with the likes of you.	
Slippery Slope	One thing leads to another, and before you know it. . . .	We should never abandon the tried-and-true methods and experiment with new technologies. If they don't work, we risk losing to competitors, squandering funds, and eventually bankrupting the company.	The final conclusion is not a necessary outcome of the initial assumption. There is no logical chain of events.
Slippery Slope		If we permit casual attire on Fridays, the work ethic will deteriorate, and pretty soon people won't show up for work on Fridays at all.	
Complex (loaded question)	Two unrelated issues are joined in the same question—either way you answer, you lose.	Has your company stopped mistreating its employees?	If you answer "yes," you admit that you were mistreating them in the past; if you answer "no," you are still mistreating them. Show that these are two illegitimately joined questions.
Changing the Subject		Yes, but what about. . . .	
Self-Righteousness		But I spent all night preparing for this presentation. You must give me the account.	Real life awards results, not efforts.
Reasoning/Inductive			
Hasty Generalization		In six of the bridge collapses in the last five years, faulty welds were identified as the cause of structural failure. Therefore, faulty welding is the major cause of bridge collapses.	Six failures of how many total? What were the other causes? Are conclusions based on a representative sample? See the Statistical Lies discussion.
Unrepresentative Sample		People representing Company A all look bright, enthusiastic, and competent. Let's award them the contract.	The dull, unenthusiastic, and incompetent were not brought to the presentation. And how can you know that those that look competent actually are? Check the credentials.
Slothful Induction		I know we weren't successful on the last twenty-seven bids, but I just have a feeling that we will get this one.	What is the basis for your argument if the evidence or precedent is to the contrary?
Reasoning/Statistical Syllogisms			
Accident or No Exceptions		This is the way it's been done here for years. The corporate policy states that you must have a requisition form to order your critical parts. We're out of forms, so you can't have your parts.	Show that it is contrary to common sense to follow rules, even when they do not apply in a particular circumstance, just because they exist. Every rule must have exceptions.
Converse Accident or Inappropriate Exception		Because we allowed Customer A, whose plant burned down last week, to pay late, we're going to allow all our customers to pay late.	Exceptions are just that—and cannot be extended to all cases.
Reasoning/Causal			
Post Hoc	After this; therefore, because of this. . . .	We brought in a new senior executive vice president of special projects, and two weeks later, our losses decreased by 50 percent.	Do not confuse correlation with causation. Just because two events occur concurrently does not necessarily mean that they are related to each other—i.e., one does not necessarily cause the other. An apparent relationship can be established between any two events or variables. Some correlations can be fascinating but totally irrelevant from the causation standpoint.
Joint Effect		We're having a number of key people resign as a result of customer dissatisfaction.	Whereas, both the dissatisfaction and resignations may be caused by the sloppy service we provide.
Complex Cause		The accident was caused by the wrong location of the electrical pole.	True, but it may not have occurred had the driver not been drunk and speeding.
Genuine but Insignificant Cause		We must cut costs. Every paperclip counts.	
Reasoning/*Non Sequitur*			
Affirming the Consequent		If our competition were stealing our employees, we'd have a high turnover. We do indeed have a high turnover. Thus, the competition must be stealing our people.	Couldn't these individuals just be leaving on their own? Even if the premise is true, the conclusion can be false.
Denying the Antecedent		Breakdown in the production line will cause delays. But a breakdown did not occur. Thus, there will be no production delays.	Of course, breakdowns are not the only causes of delays.

Table 3. Identifying and Refuting Major Fallacies

Following is my own interpretation of the four types of doublespeak, as applicable to data analysis in crisis management, and as doublespeak deteriorates into doublethink.

Euphemisms are code words used to sugarcoat unpleasant realities into more palatable situations. Often they are used to diminish the seriousness of the situation by understating its importance—thus the understatement. The existence of euphemisms and understatements and their masterful misuse distorts the perception of reality and makes it possible to have "negative cash flow" instead of losses, "friendly fire" instead of killing our own troops, and "strategic redeployment" instead of retreats—not to mention the various disguise words for layoffs or firings, including but not limited to restructuring, reengineering, downsizing, rightsizing, etc., ad nauseam.

Jargon is the overuse and misuse of technical and professional lingo by the "experts" when addressing the general public. It is designed to make things look more, not less, complicated. It goes beyond the patronizing oversimplification by implying that we, the unwashed layman masses, just do not have the mental capacity to understand the issues at hand and should leave the judgment to the experts.

Gobbledygook or *bureaucratese* is simply incoherent mumbo jumbo.

Inflated language is designed to embellish things and give importance where none is due. It is the opposite of understatement. Job titles are an excellent example—and a cheap substitute for higher pay, we may add.

Doublespeak is particularly dangerous to the clarity of the mind, freedom of thought, and pursuit of the truth because, given enough time and repetition, it numbs us into actually believing the nonsense—or at least stops us from challenging it intellectually. Insipid doublespeak thrives in our currently fashionable, politically correct times. It allows us to become talking heads and speak without saying anything, meaning anything, or offending anyone. We listen and know what the other individual wants to say, but she can't, so that we sympathize with her and translate silently. Soon thereafter, we stop translating. We start rationalizing to ourselves that which we know is not true, in order to alleviate the unease of self-doubt. After all, it's much more comforting to be in a midst of a mild period of negative growth than in an outright crisis, regardless of the reality.

Doublespeak always carries with it the fear of repercussions. We dare not speak against the incompetent boss for fear of getting fired. But, our boss does not want us to fear him, nor to fire us. All he wants is for us to agree with him, repent our transgressions, and truly accept his incompetent ways. In the beginning, we still see the errors and blunders, and they bother us. After a while, however, with long-term repetitive dosages of doublespeak, our brains become lobotomized into the Orwellian world of doublethink—we have now been successfully immunized against the ability to think for ourselves. As distasteful as it is to paraphrase a Nazi propagandist, it is applicable as a warning that we must not surrender our freedom to think clearly and acknowledge reality in our business, political, and personal lives. It was Joseph Goebbels who said, in effect, that *people believe that which they think is true, and they believe to be true what they hear the most.*

Doublethink is the worst cognitive trap of all. If you have fallen into it, you will self-delusionally slide through your crisis straight into bankruptcy while honestly believing to the bitter end—and even after the end—that you are steadfastly on the right path. Doublethink allows us to express belief in two conflicting positions with a straight face—to feel strongly both ways, while not quite becoming schizophrenic.

Why? Doublethink allows us to internally reconcile cognitive dissonance—at least temporarily. We stop questioning and start blindly trusting what we are told by the "authorities." Surrendering our reasoning ability to the "experts" is comforting—it frees us from the burden of responsibility required for independent thinking, decision-making, and action. Doublethink is the getaway car for our *Escape from Freedom*, to borrow the title from Erich Fromm's classic book. We slip into a compliant, childlike state of mental codependency.

In spite of all efforts to the contrary, perceptions cannot hide reality forever. Cognitive dissonance—the difference between what we are forced to believe and what we know is true—is most often resolved suddenly. Reality is merciless—it is what it is, not what we wish it to be. Unfortunately, institutionalized corporate doublethink can mask the reality for a long time.

Perhaps the epitome of understatement and doublespeak is the Japanese emperor's speech to the nation at the end of World War II, explaining Japan's unconditional surrender with the statement that "the war has not necessarily progressed in our favor." It is a classic example of doublespeak exemplifying doublethink to the end—with a very violent reconciliation between perception and reality, we might add.

Learn to recognize logical fallacies and doublespeak, then eliminate; cut through the bull to the substance.

Analyze Numerical Data

Analyzing numerical data, as opposed to verbal data, requires a change in thinking from intuitive to mathematical. Very few people are comfortable with expressing themselves both verbally and numerically; it is therefore important to cultivate the weaker side and seek advice when needed. From the sequence standpoint, it is preferable to eliminate verbal inconsistencies first; thus, you will not be influenced by the nuance and style in which the hard data is presented. The following numerical analysis steps are listed in continuation with the verbal analysis.

Step 6—Eliminate False/Irrelevant Trends

When analyzing data flow over time, it is essential that the data truly represent the essence of the trend. For example, look at the data table in Figure 25 and the three graphs, all derived from this data, in Figures 26, 27, and 28. The first graph (see Figure 26) is a composite of all data types; variations from period to period appear totally randomly. The second graph (see Figure 27) shows defragmented data: base trend with overlay cyclical, seasonal, and other components. When the overlays are removed, we get a "clean" linear data trend in the third graph (see Figure 28). We have filtered out those "overlays" that do not directly exemplify the data.

PERIOD	1	2	3	4	5	6	7	8	9	10	11	12
Base Trend	2	4	6	8	10	12	14	16	18	20	22	24
Seasonal Component	2	4	6	8	6	4	2	4	6	8	6	4
Cyclical Component	1	2	3	2	1	2	3	2	1	2	3	2
Random Component	1	–5	4	–7	2	–12	4	–9	9	–20	2	3
TOTAL	**7**	**7**	**22**	**15**	**24**	**12**	**30**	**21**	**43**	**20**	**44**	**45**

Figure 25. Trend Data Table

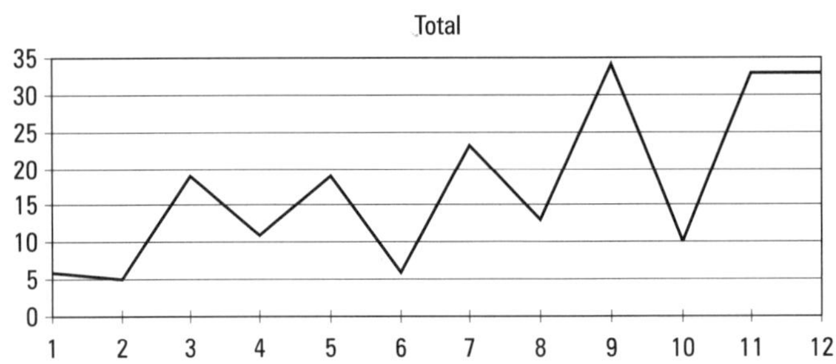

Figure 26. Composite Data Trend

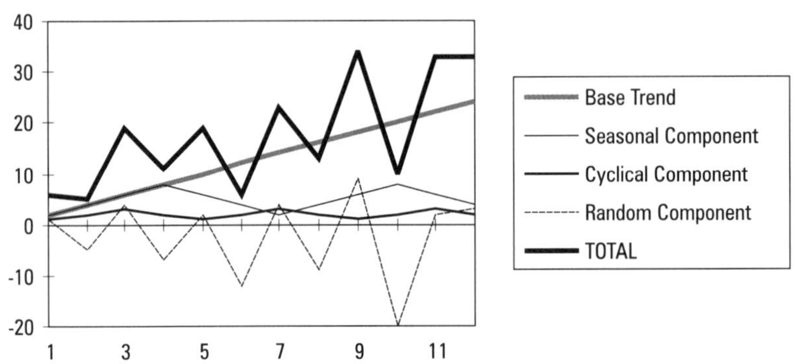

Figure 27. Defragmented Data Trends

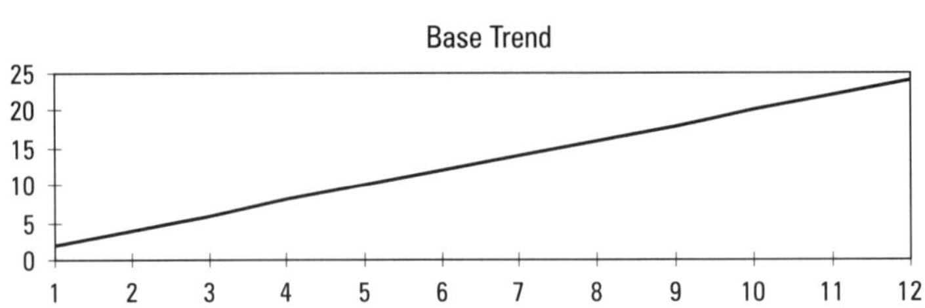

Figure 28. Base Data Trend

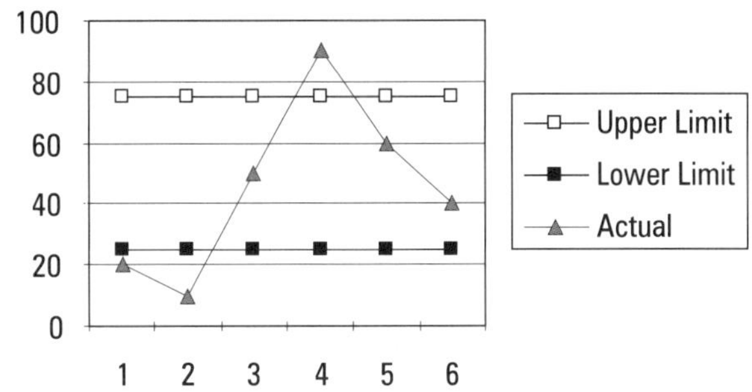

Figure 29. Acceptable Data Range

Step 7—Eliminate Statistical Lies

For every credibility gap, there is a gullibility fill.

Sample? What sample? Believe it or not, data backup can be contrived statistically to represent any desired conclusion. All we have to do is limit our data collection to a sample that supports our preconceived notions, and steer judgment toward forgone conclusions.

This limited or "stratified" sample must be representative of the larger cross-section of data in order to be truly representative. Otherwise, it will be biased and steer judgment toward preconceived conclusions, which is often the favorite tool of seasoned spin-doctors. For example, if you wanted to "substantiate" the prevalence of financial fraud, your sample would include the incarcerated white-collar criminal population.

This concept is closely related to the inductive fallacies mentioned in the previous section. Inductive reasoning depends on the ability to draw inferences or to extrapolate, based on the similarity of a sample. The more similar the sample is to the overall population, the more valid the inductive inference.

An alternative is random sampling; it is reliable but expensive, since each item must be checked.

Average? What average? An average can be arithmetic, weighted arithmetic, median, mode, each representing reality with a different twist.

Statistical average does not represent the majority of the cases; statistically, the average size of a company in your city may be 500 people, or $3 million in sales, but there are more companies with fewer (or more) people than 500. Or, consider this statement: *according to an American Management Association survey, people make a correct decision about 50 percent of the time.*

Here are some ways to guard against statistical lies:

- ✔ Establish a real comparative basis—more or less than what?
- ✔ Get the comparative rate, not an absolute number—e.g., fatalities per million passenger miles.
- ✔ Differentiate between change versus the rate of change.
- ✔ Establish an acceptable or normal range, with upper and lower limits, rather than just one "correct" number, as per Figure 29.

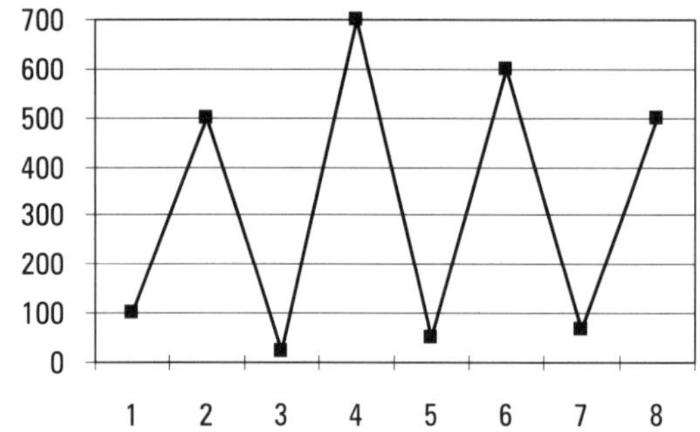

Figure 30. "Erratic" Data

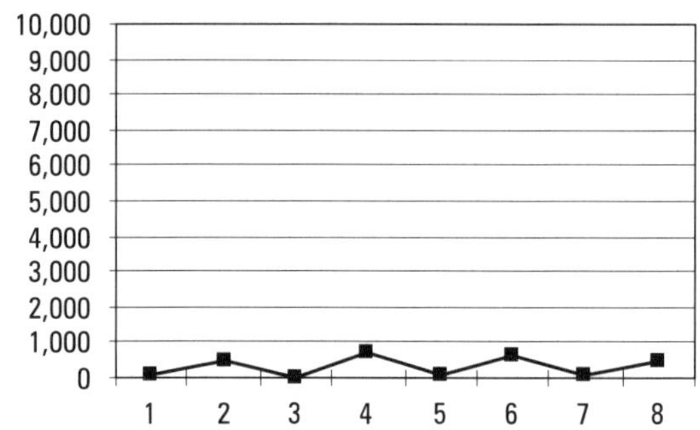

Figure 31. "Low" Data

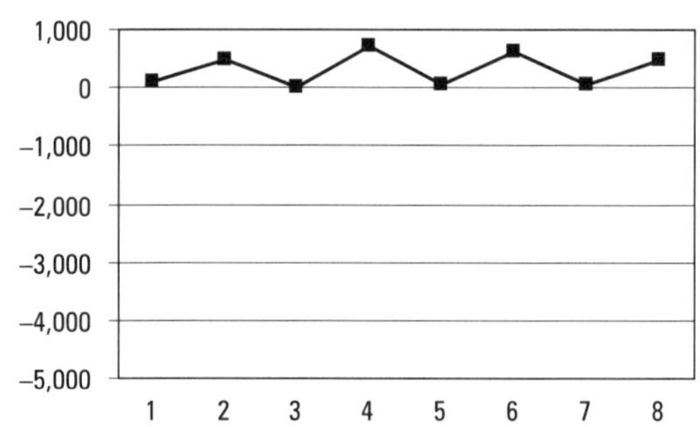

Figure 32. "High" Data

Data Manipulation—Statisticulation

Here are several ways to (mis)represent the same set of data, depending on the particular set of circumstances, motives, or audience.

Distorted Frame of Reference

Changing the scale, frame of reference (see Figure 30).

Is the trend too erratic? Need to show less fluctuation and more stability with profits, employee turnover, for example? No problem; reduce the scale (see Figure 31).

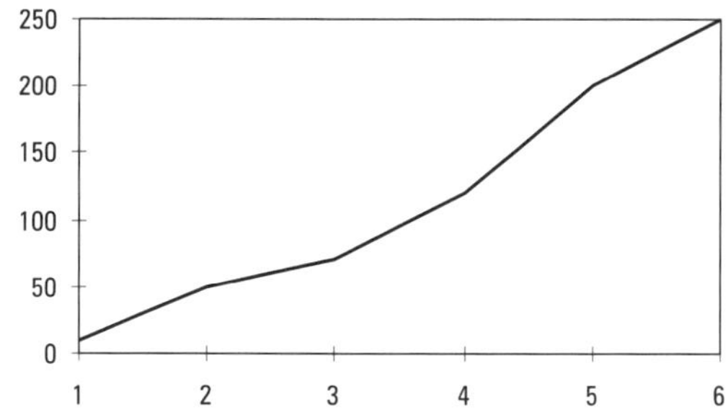

Figure 33. "Steady" Growth

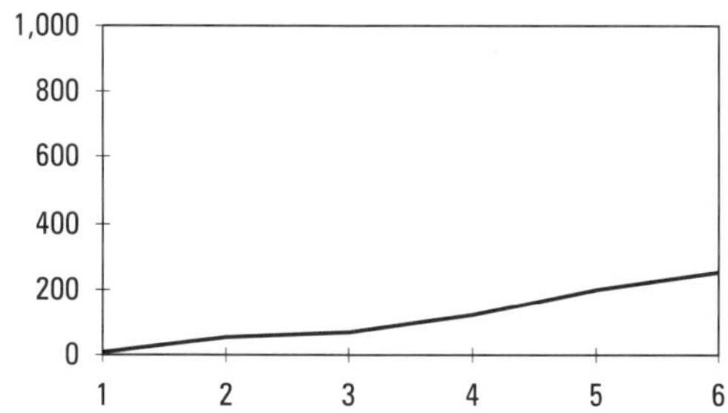

Figure 34. "Slow" Growth

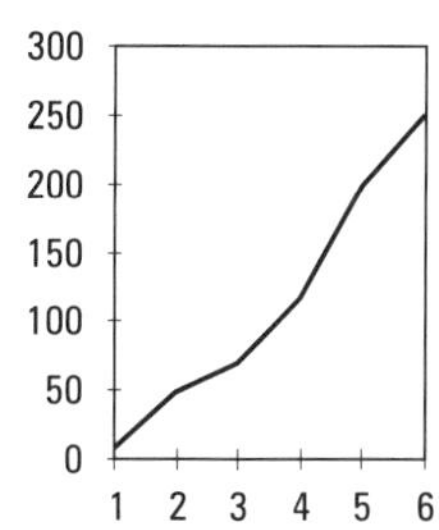

Figure 35. "Fast" Growth

Now, does the data seem too "low?" No problem; shift the data to the top of the scale, then change the scale to show how you are above the "minuses" (see Figure 32).

You can reverse the procedure if the data seems too "high."

In all three graphs (Figures 30, 31, and 32), identical data are presented differently and can therefore be distorted and interpreted differently by an uncritical observer.

Learn to view facts as facts, separate from the context or form in which they are presented.

Distorted Aspect Ratio/Scale

By changing the aspect ratio (between the x and y axes) and the value scales in the graphs in Figures 33, 34, and 35, we can make the same set of data "look" different. The graph in Figure 33 represents a steady growth

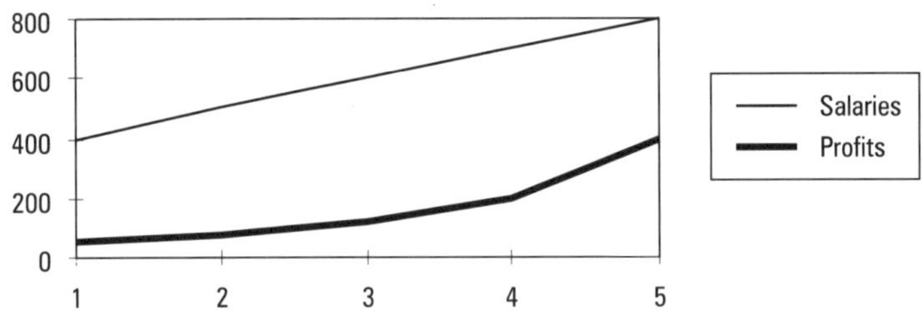

Figure 36. Absolute Change

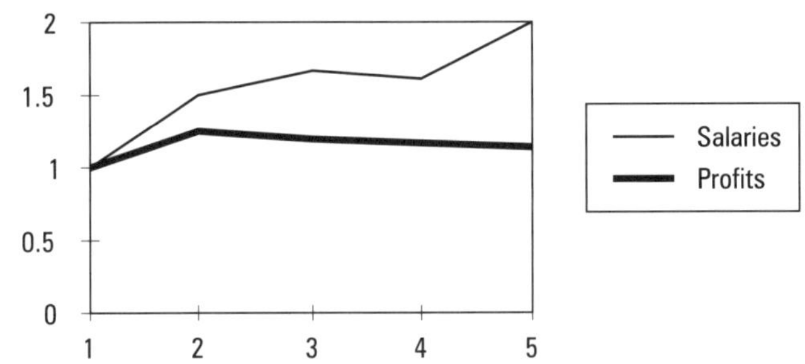

Figure 37. Relative Change

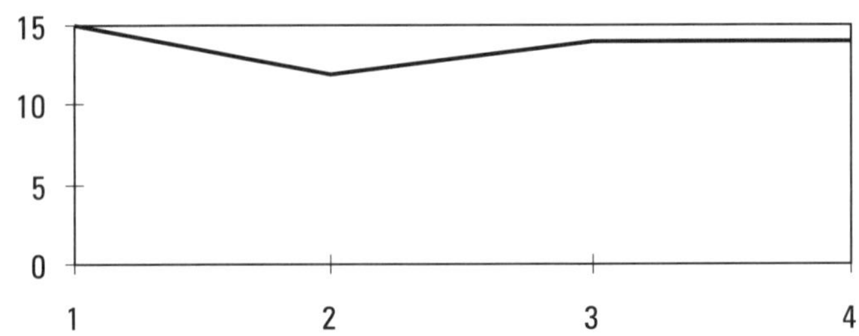

Figure 38. Quarterly Profits

trend—but growth of what? If it is a negative trend—say, debt increase—we may want to use the graph shown in Figure 34. If it is a positive trend such as a rise in profits, we'll use the graph shown in Figure 35.

Mixing Absolute with Relative Change

Look at the graphs in Figures 36 and 37. Did the salaries rise faster or did the profits? What very different message does each graph send? Note that the graph in Figure 37 shows the rate of change compared to the previous period.

Skipping Interim Data

Missed readings in between can be significant. If we only look at the quarterly profit graph (see Figure 38), the performance seems consistent enough; but a quick look at the monthly figures (see Figure 39) shows a wide fluctuation.

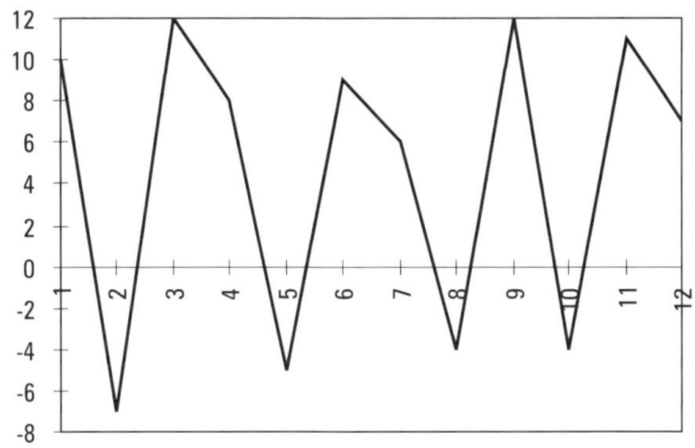

Figure 39. Monthly Profits

This brings us to the conclusion of the data-analysis process. After filtering the mass of data through the analytical steps, you should be able to weed out all the irrelevant and incorrect information, and be left with a bare factual minimum required to proceed to the decision-making stage.

Decision-Making

When you see a fork in the road, take it.

Yogi Berra

The objective of the diagnostic process, data collection, and analysis is to determine the one best solution to the problem. Ideally, the decision-making process is then concerned not with the evaluation and selection of the alternatives, but with the *implementation* of the solution, rather than trying to intellectualize the unpredictable with "decision theories."

Forecasting, scenario development, simulation, and other crystal-ball techniques are prediction models and require subjective evaluation. Combinations of probabilities and outcomes are endless, and attempting to predict or simulate each imaginable circumstance is futile.

You can't be prepared for all contingencies verbatim. It is much better to cultivate rational, fact-based, decision-making habits that will give you the innovative flexibility of mind to react in a crisis situation. In his book, *A Man On the Moon*, Andrew Chaikin points out that NASA does not simulate multiple and unrelated failures in spacecraft, since failure's combinations are infinite but probability of occurrence, infinitesimal. Rather pilots and astronauts are taught reflexive, ad hoc decision skills. Therefore, instead of an in-depth analysis of the formal decision theory, we will focus on the pitfalls and obstacles to a rational decision-making and implementation process.

Let us first identify the conditions needed to make a rational decision. We will look at the characteristics of an irrational decision process later.

Required Conditions for Making a Rational Decision

1. Problem has been identified and cause diagnosed. Failure to understand cause almost always causes the errors to be repeated, usually with progressively worse results.
2. Required data has been collected, analyzed, and converted to facts.
3. Accurate diagnosis of the problem is complete: You must have an accurate picture of both the reality and the desired state of affairs.

Decision Classification

There are two basic classifications of decisions:

➔ Need versus opportunity—are we forced to make a decision, or do we have an opportunity to do so?

➔ Generic versus unique—is it a standard or a custom solution?

We will address the latter classification first, while the need versus opportunity classification will be discussed here.

Four Types of Decisions by Generic versus Unique

In Part III, Crisis Prevention, we will talk in detail about the importance of standardizing routine operating procedures and decisions, so that we can allocate attention to the truly serious and important unique problems. Consider the following four decision types in ascending order of "uniqueness."

Generic: Adaptations of Routine Processes

One size fits all; you can be pretty sure that the decision you have made yesterday can be repeated today for the same type of routine, mundane situations. Examples include processes such as inventory, production, and logistics.

Unique for One Case, but Generic for the Type

Each decision appears unique on a case-by-case basis, but you are really dealing with a generic, repetitive situation. Take, for example, construction projects or mergers and acquisitions. You can apply standard decisions after customizing for idiosyncrasies such as personalities (bankers, clients, CEOs, architects), corporate "culture," etc.

Early Manifestation of a Generic Problem

At first it appears that there is only one unique problem case, but soon many other similar cases appear. Initially unique decisions can later be standardized. A good example is reacting to an outbreak of a new virus.

Truly Exceptional Decisions

As the name implies, unique problems require one-of-a kind custom decisions.

Fighting Fires

Decision effectiveness and decisiveness problems most often occur when we treat generic situations as unique. We fight fires instead of applying standard decisions to routine problems, essentially reinventing the wheel for every new case of a generic problem thrown at us. This concept ties into the triage method, discussed earlier.

To stop reinventing the wheel, we need to codify the knowledge gained from past decisions into rules of thumb that can be applied in similar future situations. The process of implementing rules of thumb or shortcuts to expedite generic decisions and avoid fire fighting is known as "heuristics."

To summarize:

- Fire fighting is treating a generic situation as if it were a series of unique problems.
- To the contrary, it is a mistake to treat a new event as if it were another example of an old problem—unique problems require unique solutions.

Next, let us briefly review the idealized versus real-world conditions under which decisions are made.

Idealized Decision-Making

→ Assumes decision-maker takes into account *all* relevant information.
→ Assumes decision-maker integrates relevant information accurately.

Real-World Decision-Making

> We specialize in precision guesstimating.
>
> Coffee Mug Label

→ Decision-makers do not always take into account all relevant information.
→ Decision-makers do not always integrate relevant information accurately.
→ Decision-makers regularly deviate from rational decision-making—they are influenced by decision biases.

Rational Decision Process

> Be always sure you are right, then go ahead.
>
> Davy Crockett

1. Establish the need for a decision.
2. Define decision objectives and deadlines.
3. Establish and weigh decision criteria.
4. Generate alternatives.
5. Evaluate consequences and impacts.
6. Rate and select alternative.
7. Test/model.
8. Implement.

Establish the Need for a Decision

First, we must be clear about the reason for making a decision. Every time we have a choice to select from among different alternatives, we have an opportunity to make a decision. Sometimes choices are forced upon us, or they just fall into our lap; at other times, we actively pursue them.

You should make a decision if:

- Not making a decision is clearly the worst alternative, and the current state of affairs is untenable—this is a *crisis decision*.

> A bad decision is better than no decision.
>
> Adm. Wm Halsey

- You are forced to resolve a problem by selecting a course of action—this is a *problem decision*.
- An opportunity presents itself, and you can positively influence the outcome—this is an *opportunity decision*.

Define Decision Objectives and Deadlines

What are you trying to achieve by making a decision? If the objectives are not defined, then you could be solving the wrong problem, and you could end with a perfect solution to an irrelevant objective.

Work is long, and time is short. In a crisis situation, you must make a minimum number of crucial decisions—as few as possible and as important as possible. Lowest common denominator applies:

✔ Minimum amount of data required to establish facts.
✔ Minimum number of facts required to make a decision.
✔ Minimum number of decisions required to resolve the crisis.

Let us say that you are a factory manager faced with mounting operating losses, decreasing productivity, production and shipping delays, and increasing employee turnover. These are the symptoms of a crisis. Based on an extensive data analysis, you determine that the main cause of the crisis is your obsolete, poorly located, and overcrowded facility, and that the problem will be resolved by adding 100,000 square feet of shop floor space and new production equipment.

The objectives that have to be met are those that eliminate the symptoms of the crisis by addressing its cause:

- Stem financial losses.
- Increase productivity.
- Reduce employee turnover.
- Reduce production delays.

Establish and Weigh Decision Criteria

Next, identify the specific criteria that must be met to address the cause of the crisis:

- Cost—deliver the new space at a minimum cost.
- Schedule—deliver the new space in the shortest time possible.
- Quality—provide high-quality space to facilitate production equipment efficiency.
- Accessibility—provide improved access to modes of transport.

Are all of these of equal importance? In idealized decision-making, probably yes; you want to have the best-situated, highest-quality space at the lowest possible cost and in the shortest possible time. In reality, not all of these criteria can be met equally well. Something has to give. For example, cheap land to build may only be available at a distance from major roads. Thus, you have to rank your criteria according to importance. You must satisfy the criteria in descending order of importance with a risk that less important criteria will have to be sacrificed.

Criteria can be mutually complementary or conflicting. Again, under ideal conditions, complementary criteria are preferable; if we satisfy one criterion, then the others are also positively influenced. In real-world crises, the criteria are mostly conflicting. In order to shorten the schedule, we must spend more money. What is more important: time or money? Or to rephrase, how much money are we willing to spend to gain one unit of time. This principle is called "comparison on an equalized basis" and is the foundation of decision tradeoffs.

The only way to establish an equalized comparative basis is to convert all the criteria to a common currency. This is essential for an "apples-to-apples" evaluation of alternatives. The best common currency, of course, is money. We can quantify that bringing the new facility online one month earlier will save $1 million. Thus, one month of time is "worth," and can be exchanged for, $1 million. But how do we evaluate subjective, "soft" criteria such as "accessibility." Again, by determining how much premium we would be willing to pay for a "better" access, or conversely, how much money we would be willing to sacrifice for a "poorer" access. If the more expensive facility alternative also has worse accessibility, the decision should be a no-brainer. But that is seldom the case; you get what you pay for. In the, end you can be analytical and quantify.

You may have several different options to meet the stated criteria. These are your alternatives. Each alternative course of action will have a set of constraints impeding the achievement of your objectives.

Generate Alternatives

- ✔ Break multiple-choice questions into default yes/no alternatives.
- ✔ Reduce decision complexity to the lowest common denominator—make it an *or* decision, not an *and* decision.
- ✔ Avoid combination of alternatives—e.g., the selection should never be "A and B" or "All of the Above."
- ✔ Establish clear-cut consequences for each decision: What happens for each go/no-go choice?
- ✔ Alternatives have to be equalized, "apples to apples."
- ✔ If unavoidable, same decision biases must apply.
- ✔ Each alternative must answer simple, not complex, questions.
- ✔ Evaluate consequences based on relevant decision factors/objectives.

Example: Complex (multiple-factor) alternatives:

- Expand existing facility.
- Move and build new facility.
- Move and lease new facility.
- Remodel existing facility.

Consequences of the alternatives cannot be compared on an equal basis.

This is a complex situation involving several basic decisions and their combinations. Break the complex decision into a set of either/or simple decisions:

Move or stay.
And:
Build or lease.
And:
Build new or remodel.

- Stay
  - Expand
  - Remodel
- Move
  - Build
  - Lease

These decisions must be made in a proper sequence. First, make a base decision—the one necessary for subsequent decisions. You cannot decide to build a new facility and then decide also to stay if there is no space to build a new facility.

Evaluate Consequences and Impacts

Consequences will generate new alternatives. These new alternatives may or may not be feasible, based on the established objectives, decision criteria, and constraints. You should always have a fallback position—Plan B.

Rate and Select Alternative

Expected value (EV) is the income (I) that an alternative would produce when multiplied by its probability (P). Decision-makers can compare alternatives and choose the highest EV.

➔ Alternative = a choice that we make.
➔ Consequence = outcome(s) of that choice.
➔ Payoff algorithm: EV= I x P%.

Once you have considered all the consequences and evaluated your alternatives according to the selected criteria, you are ready to select the best alternative. This is as far as the analytical decision-making models and tools discussed in the next section can take you. Actual implementation of a decision often requires dealing with more ambiguous and subjective constraints and obstacles.

Test/Model

Time and logistics permitting, testing a solution is cheaper than a full-blown implementation.

Implement

I used to be indecisive, but now I'm not so sure.

Here comes the hard part. You will have to surmount objective and subjective obstacles as you select your alternative and put it into action. The remainder of this chapter is devoted to overcoming these obstacles.

Alternatives	Decision Criteria	1 Stay & Remodel	2 Stay & Expand	3 Move & Build	4 Move & Lease	Value per Unit
Alternatives	Cost	($3,000,000)	($4,000,000)	($7,000,000)	($9,000,000)	
	Schedule	–6	–8	–12	–2	$500,000
	Access	D	D	B	A	$2,000,000
	Quality	D	C	A	B	$500,000
Equalized	Cost	($3,000,000)	($4,000,000)	($7,000,000)	($9,000,000)	
Payoff	Schedule	($3,000,000)	($4,000,000)	($6,000,000)	($1,000,000)	
	Access	$2,000,000	$2,000,000	$6,000,000	$8,000,000	
	Quality	$500,000	$1,000,000	$2,000,000	$1,500,000	
Equalized	Total	($3,500,000)	($5,000,000)	($5,000,000)	($500,000)	($500,000)

Figure 40. Equalized Cost-Benefit/Payoff Matrix

Decision-Making Models and Tools

We will briefly review two analytical decision-making tools used in operations research and management science: payoff matrices and decision trees.

A payoff matrix is a tabular format for evaluating a number of options under a number of constraints or variables. There are several formats, depending on the type of data analyzed, desired objective, willingness to take risks, and other factors.

Let's look at the cost/benefit format in the context of the facility expansion example discussed earlier. In Step 4, we generated several alternatives, and in Steps 5 and 6, we discussed evaluating the consequences and rating the alternatives based on the established criteria: cost, schedule, access, and quality.

The problem arises when we have to grade "soft" criteria—e.g., cost and schedule can be expressed in "hard" dollars or months, but access and quality are graded subjectively, say, with letter grades. How do we compare an A in access to an A in quality?

To resolve this problem, we must convert the values of all criteria to a common currency—and most often, money is the equalizing currency. If we can establish the worth of a grade unit of other criteria in terms of dollars, we can make even, apples-to-apples tradeoffs. As you can see, one month is "worth" $500,000; thus, we would be willing to extend the construction schedule by one month in exchange for a cost reduction of $500,000. How do we convert nondollar values to dollars? By defining how much money the nondollar values cost us. For example, the lost revenue from the new facility could be $500,000. Or direct highway access versus secondary road access would save $2 million in transportation costs.

Once all nondollar values are converted to dollars, we can convert to an equalized payoff matrix. Note that the cost and schedule are something we spend to gain access and quality; thus, cost and schedule are in negative dollars and access and quality, in positive dollars. Alternative 4 has the highest total equalized payoff—i.e., the lowest equalized expense (see Figure 40).

Sell Price	10
Buy Price	5

Buy	Sell				Result
	7	**8**	**9**	**10**	
7	35	35	**35**	**35**	
8	30	40	40	40	
9	25	35	45	45	
10	**20**	**30**	40	50	
Total	**20**	**30**	**35**	**35**	**35**

1. For each possible outcome, select the *worst* possible alternative (lowest profit).
2. Select the *best* among the worst outcomes.

Result: Buy 7 and risk not being able to sell more.

Figure 41. Maximin—Minimizing the Downside

Sell Price	10
Buy Price	5

Buy	Sell				Result
	7	**8**	**9**	**10**	
7	**35**	35	35	35	
8	30	**40**	40	40	
9	25	35	**45**	45	
10	20	30	40	**50**	
Total	**35**	**40**	**45**	**50**	**50**

1. For each possible outcome, select the *best* possible alternative (highest profit).
2. Select the best among the *best* outcomes.

Result: Buy 10 and hope to sell all.

Figure 42. Maximax—Maximizing the Upside

Payoff Matrices

Next let us compare the optimistic and pessimistic payoff analyses. Maximin and maximax are commonly used terms in decision theory and operations research. *Maximin* is the pessimist's choice: you want to minimize losses. *Maximax* is the optimist's choice: you want to maximize gains. Let us say you are buying widgets at $5 each and selling them at $10 each. How many should you buy to make the most money (optimist) or to lose the least money (pessimist)? That depends on how much you can sell.

Let's limit the analysis to buying between seven and ten widgets. Note (see Figure 41 and Figure 42) that the profit amounts are identical for both the maximin and maximax matrices, but the choices are entirely different.

Alternative	Consequence		Income (I)	Probability (P)	Expected Value (EV)
Alternative 1 Sue the Contractor	Consequence 1	Win	$500,000	30%	$150,000
	Consequence 2	Win	$300,000	20%	$60,000
	Consequence 3	Lose	($30,000)	50%	($15,000)
Alternative 2 Settle Out of Court	Consequence 4	Win	$100,000	100%	$100,000

Figure 43. Decision Tree

Decision Trees

The rational decision process can be graphically represented with a widely used tool known as a decision tree (see Figure 43). Each decision point is a fork in the road with possible alternatives branching out. Each selected alternative will have a number of possible consequences. Identifying all available alternative courses of action and evaluating all possible consequences is practically impossible, but a decision tree at least provides a disciplined thought framework.

In order to construct a working decision tree, you must:

✔ Identify all possible alternatives.
✔ Predict all possible consequences for each alternative.
✔ Assign a probability for each consequence occurring.

Let's continue on our previous example and assume that you have decided to move and lease a new facility, but the contractor was a month late. You sue him for $500,000, which is the value you place on one month. The contractor is offering to settle for $100,000. You have the option to proceed with a jury trial or settle. If you go to trial, you estimate that you have a 50/50 chance of winning. If you lose, you have to pay $30,000 in legal costs. If you win, you figure you have a 30 percent chance of winning $500,000 and a 20 percent chance of winning $300,000. Of course you can spread your range, as long as the probability percentages for one alternative total 100 percent.

It is very important to note that you should be comparing expected values, not incomes. A $500,000 win is "worth" only $150,000 on an apples-to-apples comparative basis with other possible consequences, but in this case at least, it looks like you have a better chance in court.

Decision Process Constraints and Obstacles

Objective Constraints

Separating difficulty from necessity.

- Separate the need to make a decision from how difficult it is.

Time constraint.

- Solutions/actions/decisions must be timely and effective—no time for detailed analysis and collection of all relevant facts.

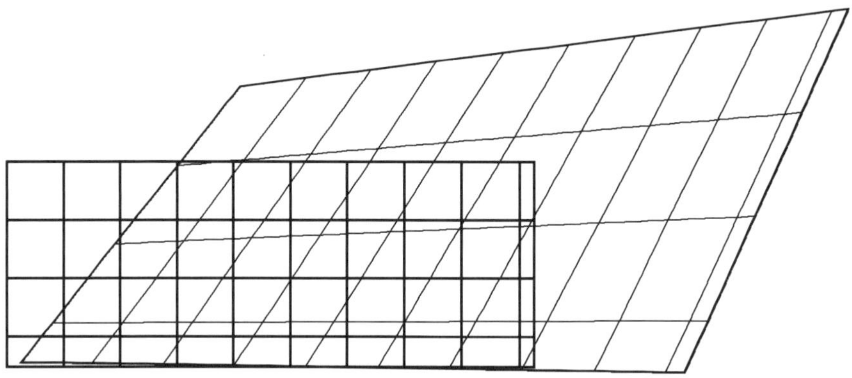

Figure 44. Factual Shift

Factual shift.

- Factual shift change occurs too fast—an analogy is a search-and-rescue mission in the ocean; water grid quadrants move, you have to adjust focus on where you anticipate finding the survivors. (See Figure 44.)

Tradeoffs and compromises.

- You can't have your cake and eat it, too.

✔ Balance/trade off between two conflicting objectives or extremes.

✔ Equalize all values to the same currency = even barter. What is X worth in terms of Y? What is it worth to you?

Uncertainty and probability. Problem with alternatives and consequences:

■ They must be *mutually exclusive*—can't have the same option in two alternatives.

■ They must be *cumulatively exhaustive*—all possibilities must be accounted for.

However, as discussed in the Decision Trees section, each decision can have multiple alternatives, and each alternative, multiple consequences—they can't all be predicted or all accounted for. Outcomes and combinations are infinite. It is human nature to try to classify, index, and organize past events in hopes of discerning trends.

This trending approach is certainly valid for many phenomena, and failure to consider the past events is a mistake. However, we have to accept the fact that some events are purely random. As the stockbrokers' disclaimers say: "Any investment involves risk. Past performance is not a guarantee for future results."

The statistics involved in decision-making under uncertainty is complex and beyond the scope of this book. For simple reference only, here are some typical trend lines and their applications (see Figure 45):

➔ **Linear**—for data increasing or decreasing at a steady rate.

➔ **Logarithmic**—for data increasing or decreasing rapidly and then leveling off/plateauing.

➔ **Exponential**—for data increasing or decreasing at an ever more rapid rate.

➔ **Polynomial**—for fluctuating data with ups and downs.

Note that we have several different trend lines based on the same set of data to illustrate that each projects the future differently. If the actual data in

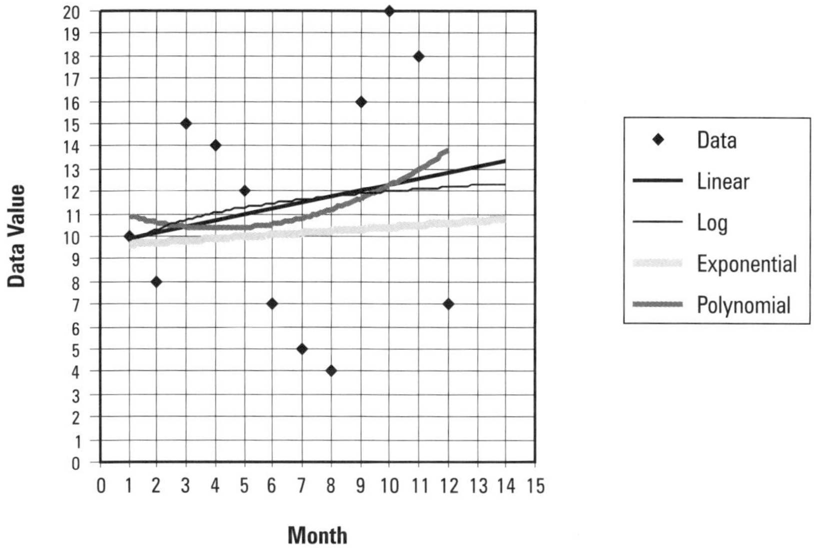

Figure 45. Data Trends

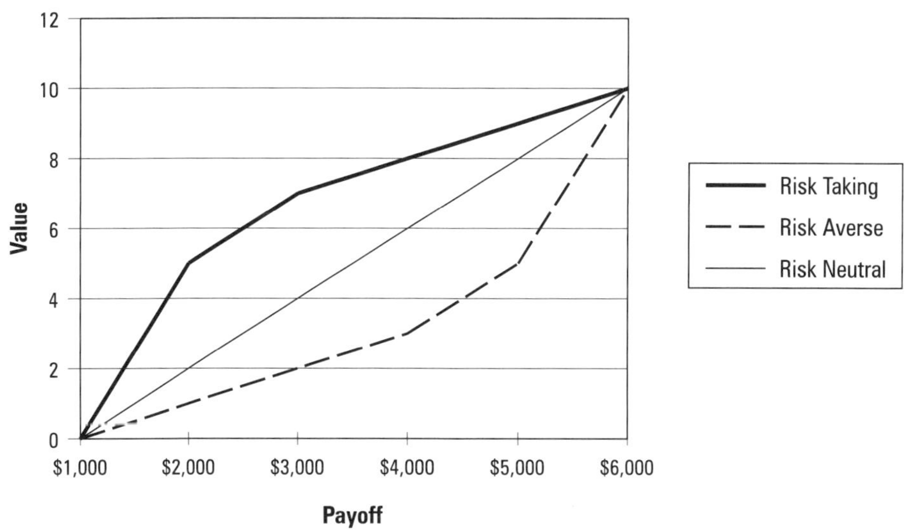

Figure 46. Risk Profile

the future periods is different from the projected, as it usually is, it changes the overall trend line—we didn't get the future right, but at least we have the 20/20 hindsight to explain how it all happened.

Risk mitigation. Are you a risk-taker or are you risk averse? What are the ways to minimize risk? Methods include ensuring the downside, spreading and sharing of the risk, etc. Figure 46 illustrates the range of decision options between risk-taking and risk aversion by assigning the value one places on the probability of success. If 10 is the highest value, and you are ensured of a 100 percent payback, then there is no risk; anyone will place a value of 10 with a

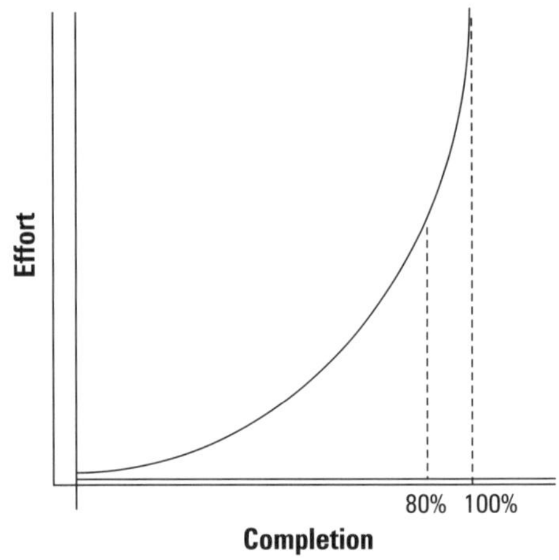

Figure 47. The Perfection Trap

100 percent chance of success. But, how much is a 50 percent chance worth to you? Maybe 7 if you are a risk-taker, but only 3 if you are risk averse.

Consequences. As discussed earlier, all alternatives yield consequences, positive or negative. It is equally detrimental to the rational decision-making process to either exaggerate or ignore the importance of each consequence.

Impacts. Impacts are "down-the-line" consequences that may affect future decisions linked to the current decision. How will the decision that you just made influence the decisions that you will have to make next month?

Subjective Constraints

Paralysis by analysis.

- Decisions must be made based on the least sufficient amount of background data required for the solution. By the time you analyze all the available data, the window of opportunity to resolve the crisis will be gone, and the crisis will, in the meantime, progress to failure.

  The perfection trap.

> The pursuit of excellence is gratifying and healthy; the pursuit of perfectionism frustrating, neurotic, and a terrible waste of time.
>
> Anonymous

- The futility of perfection is that gathering the last 20 percent of the data required to make the "perfect" decision will take 80 percent of the time (see Figure 47). In the meantime, a significant amount of gathered facts will become obsolete (reference to Pareto rule discussed in Data Collection section).

Shooting from the hip. Don't go too fast or act rashly. Don't be afraid to do nothing—if that is the best choice, or if your triage procedure tells you to allocate your resources to other priorities.

Groupthink.

> I don't want any yes-men around me. I want people to tell me what they think even if it costs them their jobs.
>
> Samuel Goldwyn

Groupthink is decisions designed by a committee. Here are some of the symptoms of groupthink to watch for:

- Decision-making by committee/jury.
- Wasted time and slower decision-making.
- Illusion of invincibility/righteousness/strength in numbers.
- Forced consensus and conformity/ostracism of dissenters.
- Domination and goal displacement.
- Internal amplification/validation/reassurance/backslapping—like lemmings off a cliff.
- Downplaying the problems.
- Ignoring unfavorable evidence (also see Decision Biases or Traps, the following section).
- Unwillingness to admit errors/cut losses (also see Decision Biases or Traps).
- Reduction to lowest common denominator—understood and accepted by everyone (Parkinson's Syndrome—institutionalized mediocrity).
- Reinforcing preconceived notions/rubberstamping.
- Absolution from personal responsibility or conscience: "just following orders" or "I was given bad advice by my trusted advisers."
- Masks individual insecurities—everybody is a "yes man."

Irrational decision-making. Alternatives to rationality include:

- Precedent.
- Voting.
- Divine guidance.
- Intuition.

Symptoms of irrational, disorganized decision-making include:

- Forced consensus—groups; no single decision-maker/committees.
- Based on polls, delusions, wishful thinking—not facts.
- Solutions proposed where problems don't exist—"compensating by raising issues"/fixing the unbroken.
- Choices made without solving the problem—flipping the coin.

Bounded rationality. Often, the demands for making a competent decision are just too much for an average human decision-maker. Humans have bounded, not infinite, rationality, and use various mental crutches to muddle through a decision-making process:

- Satisficing—doing just good enough, rather than maximizing.
- Sequential search—figuring one thing at a time.
- Simplified models of real world—real world is an information overload of a Leonardo painting; we draw stick figures so that we can understand the world.
- Precedent/performance programs/reducing complexity.
- Uncertainty absorption.

Decision Biases or Traps

- Decision-making biases are closely related to logical fallacies in that they distort reality.
- Logical fallacies are obstacles imposed by external sources in converting data to facts.
- Decision biases are subjective obstacles in using the facts to make a decision.

See Table 4 for some typical decision biases, their characteristics, and defenses in a matrix format. As in fallacies, numerous classifications and definitions exist in professional literature. As an exercise in critical judgment, try filling in the blanks.

Preventing Biases

✔ Eliminate interpretation biases and perception shifts: People misinterpret data according to their own sets of preconceived notions, filter through their value systems—facts must be completely objective.

✔ Eliminate delusions/wishful thinking.

✔ Practice rational decision-making: Always remember to maximize gain and minimize loss.

Decision-Implementation Obstacles

- Implementation obstacles should not be a concern in a crisis situation. As a Project Surgeon, you must practice business, not politics.
- Remember your role: If you are the operational manager in charge, your decision is the order to be executed, and you take responsibility for the results. If you are the consultant, propose the best decision to the inside decision-maker; if it doesn't get implemented, it's not your problem.
- Getting an organization to foster decision-making and implementation is further discussed in the Organizational Design section—unity of command and action, certainty of decisions, etc.

Political Routines

The earlier-mentioned points sound good, but the implementation of decisions in any organization usually requires "political" routines. Unless you are an absolute ruler of your corporate universe and can afford to govern by decree, implementing even a best decision is never a straightforward process in a corporate organization. There are always rules and regulations to be upheld and forms to be completed. All decisions must be validated against formal rules and informal "values." Sometimes these are valid checks and balances; more often they are obstacles planted by the self-appointed corporate guardians who justify their existence by raising issues.

Interrupt/Stall/Avoidance Techniques

- External authorizations and bureaucracy—"We do not have an approval from the Corporate Department of Paperwork."
- New option—the "what-if" delay ("but wait, that's not all").
- Scheduling and timing delays—"Come back tomorrow."
- Hurry up and wait—"I must have this report on my desk Monday morning. ... " and then it's going to sit there for two weeks.
- Feedback delays—"I'll get back to you (maybe, someday)."
- Comprehension cycles—"Explain it to me one more time, slowly."

Decision Biases and Symptoms	Example	Defense
Availability Bias		
Decision-making is unduly influenced by related information from memory.	I remember steel plate going for $700 a ton last year, so that's the figure I'm going to use for next year's material procurement budget.	Things change. Don't over-rely on memory. Get the most recent reliable data.
Biases Due to Retrievability of Instances		
Biases Due to Effectiveness of Search Set		
Anchoring Bias		
First Impression		
Adjustment and Anchoring	The asking price for the house was $250,000. I can't offer less than that without offending the seller.	Offer what you think the house is worth, not what the seller had anchored in your mind. Is the anchor price supported by market value or only by the seller's wishful thinking?
Insufficient Adjustment		
Biases in Evaluation of Conjunctive and Disjunctive Events		
Anchoring in the Assessment of Subjective Probability Distributions		
Status Quo Bias		
Steady as she goes nowhere.		
Don't rock the boat.	I'll decide how to allocate the investments tomorrow.	Tomorrow never comes.
Sunk Cost Bias		
Crying over Spilt Milk	We poured $7 million into this company. I'm sure that if we lend them $12 million more, they will start being profitable.	Maybe, but most likely not. Stop crying over spilt milk and throwing good money after bad. What's done is done. Evaluate each decision on its current pros and cons.
Hindsight Bias/Curse of Knowledge		
Delaying Regret		
Confirming Evidence Bias	Alternative A sounds better than Alternative B.	
Favorability Bias		
Ignoring Negative Evidence	Don't call me until its fixed!	
Selective Hearing	Just tell me what I want to hear.	If you have already made up your mind, look for a devil's advocate, not a supplicant who will validate your ideas.
Framing Bias		
Is the glass half empty or half full?	Let's say you have invested $1 million each in three businesses. A crisis strikes, and you have two recovery plans. Which alternative is better? Plan A will result in certain recovery of one business, and will save you $1 million. Plan B will result in a certain failure of two businesses, and will cost you $2 million.	Both alternatives offer exactly the same outcomes; they are just phrased differently. A plan sounds better because the glass is half full.
Overconfidence Bias (ego bias)		
I can't make a bad decision.		
I'm a legend in my own mind.		
Recallability Bias		
Dramatic events are remembered, even if irrelevant.		
Recent events are better remembered than the old ones.		
Base Rate Bias		
Ignoring Relevant Information		
Insensitivity to Sample Size	We'll focus the marketing effort of our luxury products division on the United Kingdom rather than on India, because Britain has more millionaires who can afford our products.	Is that so? There are twenty Indians for every Briton. Are Britons twenty times more likely than Indians to be millionaires?
Representativeness Bias		
Decision-making is influenced by stereotypes.		
Evaluating a Specific Case on the Basis of How Representative the Case Is of a Total Group		
Prudence Bias		
Erring on the Safe Side—Way Safe	Add 20 percent to the production demand estimate just to be on the safe side.	And be stuck with a 20 percent unsold inventory.
Gambler's Bias		
Misconceptions of Chance		The house always wins.
Insensitivity to Predictability		
Insensitivity to Prior Outcome Probability		
Illusion of Validity	Fractal analysis is a sure predictor of the stock market behavior.	Dont look for patterns where none exist.
Misconceptions of Regression		
Illusory Correlation		

Table 4. Decision Biases and Symptoms

- Failure recycles—"I don't think this will work; get back to the drawing board."

As you can see, getting to your decision is a rational process. Implementing it through a bureaucratic maze is not. When all else fails, try the following:

- ✔ **Bargaining**: In order to implement your solution, you may have to pit it against competing agendas in your organization. You may have to negotiate and trade certain points of your agenda in order to gain concessions from the opposing side.
- ✔ **Persuasion**: Making others see things from your vantage point and agreeing with you can only be achieved if they buy into your agenda.
- ✔ **Co-optation**: If you can't beat them, make them join you. Offer your competitors positions on your team and stakes in your agenda, but only if they abandon their own.

This brings us to the end of the Decision-Making section and to the end of Part II. By rigorously and methodically implementing the techniques learned in the first two parts of this book, the symptoms of your crisis should be addressed and successfully resolved. One major problem remains, however: the underlying conditions that have allowed the causes of your crisis to appear and grow are still intact. In Part III, we will learn how to ensure that the same type of crisis situation never occurs again.

PART III

Crisis Prevention

Corrective Surgery—Preventing Future Crises

Crisis Prevention

> The only things worth learning are the things you learn after you know it all.
>
> Harry Truman

The long-range goal of any crisis management process must be to develop permanent business mechanisms to prevent emergency and crisis situations from occurring in the future. After you have put out the fires and addressed the immediate causes, ask yourself: "What will happen to this organization next month or next year? Are all the underlying operational, systems, or human factors that caused the previous crisis situation still intact, waiting for you to go away? Is the next crisis already smoldering out of sight?"

If you are an outside gunslinger, your initial reaction may be to just keep your clients in the dark; after all, you are relying on other people's incompetence for your living—"Keep screwing up; we'll be back to fix it." However, you are not doing your job if you are helping perpetuate the human, systems, or operational factors that are the root causes of crises.

Every Project Surgeon's objective must be to work yourself out of a job—on that particular project or in that particular company where you have just finished your immediate "fixing" assignment. Your assignment is not complete until you have left the preventive management mechanisms in place. Teach your clients how to fish; don't fish for them and create codependency. Your goal is to be the Maytag repairman of business management—for one day only, however, before moving on to the next assignment.

You may argue that it is not in your interest to show your client or employer the ropes, because once she has learned your tricks, she will not need you anymore. This is, however, a short-sighted approach; you are relying on your client to continue making mistakes so that you have a built-in repeat customer cash cow.

First of all, you will be fixing the same problems over and over again—hardly a way to improve your skills. Second, your reputation will suffer if you show up at the same company, troubleshooting, year after year; people will figure that you can't be very good at what you do. Finally, and most importantly, you will eventually be exposed. Even the most dimwitted management will realize that you have been around for too long and are somehow relying on it as a captive market—are you maybe creating some of the problems just so that you can hang on?

Don't keep your clients in the dark; explain to them both what the problem is and your solution. You will earn their respect and gratitude and, contrary to initial expectations, long-term clients who will rely on your services for ever-more complex problem solving in the future.

The payoff from this long-term outlook is also greater. If you are an outsider, you can ask for a quantifiable share of the profits resulting from your system improvements or, as many turnaround CEOs do in public companies, get a load of initially worthless shares and watch their value grow as a long-range result of your work. Also, improvements don't happen overnight; remember that progress is mostly evolutionary, not revolutionary, and consists of small incremental improvements made every day. In the end, you want to point to the long-term survival resulting from your actions: "I went in and fixed the problem ten years ago—and look at them now!"

To accomplish these long-range goals, the mechanisms and procedures that you have implemented in your crisis management assignment must be made an integral part of the corporate culture and business *modus operandi*. This is a monumental task, and you will not always be in a position to introduce it or oversee its implementation, unless you are an interim or new CEO brought in to turn around the entire organization.

"Crisis proofing" the fundamental business processes and systems is the essential closure of any crisis management assignment—because contingency plans, in most cases, don't work. You can't simply leave behind an emergency checklist that will be tucked away (in a glass box with a "Break in case of emergency" sign on it) until the wings fall off. You can't foresee every possible emergency and devise a cookbook solution—this is an essential shortcoming of the case-study approach. Precedents are useful only as references. No two crisis situations are the same, and the solution that worked on a similar problem in the past may not work today. In a real crisis, the solution will not be in a manual. Rather than memorizing precedents, categorizing cases, and throwing template solutions at the problem, it is much more useful to restructure the overall corporate thinking to include a combination of prevention and vigilance in order to achieve the following twofold goal:

1. **The ability to anticipate and prevent crises: crisis prevention systems**.
2. **The ability to address and resolve those crises that do occur: flexibility, innovation, and rapid response capability**.

These two objectives can only be accomplished by bringing together the three essential business components—people, systems, and organization—and combining them in one simple formula, as follows: the right people doing the right things the right way. All three factors must be present:

- Systems (the right things).
- Organizational (the right way).
- Human (the right people).

System Component—Getting the Right Things Done

> Good judgment comes from experience, and experience comes from poor judgment.
>
> Unknown

First, design your business system to detect, prevent, address, avoid, or eliminate as many emergency and crisis situations as possible, and make the crisis management procedure the last resort, an exception reserved for the most serious problems. Remember, putting out fires should not be a way of life for a typical business manager. Crisis management in and of itself adds absolutely no value to the business process; at best, it minimizes losses and tips the scales back to zero. (Revisit the point-of-no-return concept.) Focus is on increasing the ability of your people to handle exceptions by standardizing operating procedures.

The two steps of crisis proofing the system aspect of your business are:

1. Early Detection and Warning—Crisis Radar.
2. Crisis Prevention.

Early Detection and Warning—Crisis Radar

In order to identify a crisis in the making, before the point of no return is reached, your business system must "tell" you if something is wrong. An early crisis detection and warning system must have the triggers and tripwires that set off the warning lights. The system must also "know" when something is wrong—i.e., differentiate between normal and "abnormal" situations.

Think of yourself as the captain of your business spaceship with a control panel in front of you, showing a real-time status of all critical performance parameters of your business. Your control panel—i.e., the early warning and detection system—is comprised of the following three components:

1. **Monitoring**.
2. **Reporting**.
3. **Controlling**.

Monitoring. The monitoring component consists of "gauges" that tell you the status of critical operational performance parameters of your business. Depending on your business, the gauges could be telling you the inventory levels of spare parts, accounts receivable aging, or number of customers served. Gauges can give you accurate and timely readings only if there are accurate "sensors" in the system collecting, filtering, and processing the data into facts. As discussed previously, you must have timely and accurate facts, not only for crisis management situations, but also, much more importantly, for everyday crisis prevention.

In summary, to make your business status monitoring effective, you must:

✔ Establish critical performance parameters—define what your gauges should read (the "pulse," "blood pressure," and "body temperature" equivalents of your business).

✔ Structure data "supply" to focus on the critical performance parameters only; filter the irrelevant "background noise."

✔ Think "analog"—set a scale with a minimum and a maximum on your gauge.

- ✔ Establish a range of "normal" readings—set upper and lower acceptable limits on your analog scale. Let the "system" handle the routine readings within the normal range.
- ✔ Make sure the status is monitored online/in real time. Obsolete facts are of little use, no matter how accurate.

Reporting. The reporting component consists of "triggers" or "warning lights" that notify you of abnormal readings in your gauges—breakdown in production line, overheated boiler, leak in the pipeline, delayed delivery, or other problems.

Characteristics of a good reporting system include:

- Produces verifiable facts.
- Provides timely, accurate, and relevant information.
- It is regular: allows for proactive identification and resolution of problems before they get out of hand.
- It is consistent—compared to the same set of benchmarks.
- Assigns responsibility and deadline for project resolution.
- Focuses management attention on red flags—exceptions to the rules.
- Prioritizes tasks—addresses major issues first; develops a "hot list."
- Red flags automatically initiate corrective procedure.

Controlling. The control component consists of "switches" that allow you to change or modify the parameters:

- Fly-by-wire management.
- Buttons/switches—setting a process on or off.
- Levers/dials—adjusting the speed, intensity, and relative value of a procedure.

Controlling essentially involves decision-making based on fact. Use the control mechanisms to proactively adjust, modify, change, stop/start, or increase/decrease any business process.

- ✔ Measure performance based on critical resources: time, money, and information.
- ✔ Measure performance based on established benchmarks.
- ✔ Regularly compare actual performance to plan—strive for continuous, incremental improvements.

Control methods must be:

- Measurable (against a set benchmark).
- Meaningful (measure the right things).
- Influenceable (produce desired improvement).

Controlling process involves the following:

- ✔ Change and improve standard systems and procedures.
- ✔ Build in preventive measures.
- ✔ Implement checks and balances.
- ✔ Develop action/troubleshooting checklists for future emergencies.

In most businesses, these systems do not exist—which may contribute to the crisis developing in the first place, failure to recognize it, and failure to react in a timely and appropriate manner.

Crisis Prevention

Risk magnitude is often inversely proportional to the likelihood or frequency of crisis occurrence. Take, for example, building structural failures—very rare accidents, which can have catastrophic consequences. Such catastrophic events occur when there is a convergence (all coming together at the same time) or suc-

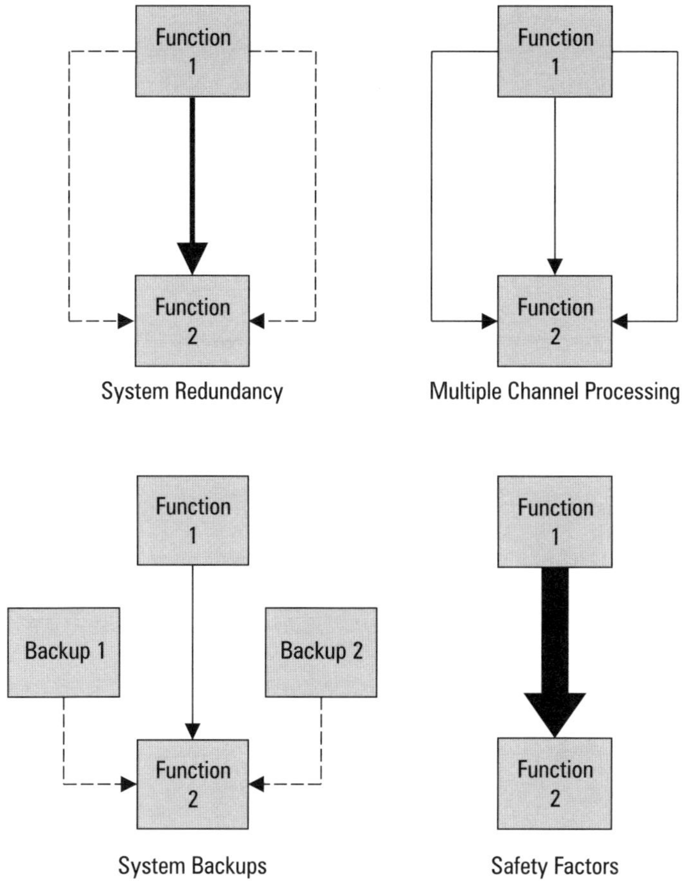

Figure 48. Crisis Prevention Mechanisms

cession (domino effect) of unaddressed high-risk factors—in other words, when all the wrong things happen at the same (wrong) time. To avoid catastrophic failures, systems—both natural and manmade—have built-in crisis prevention mechanisms, which can be categorized into five groups:

- System redundancies.
- Multiple channel processing.
- System backups.
- Safety factors (system flexibility/expandability).
- Failure modes.

Flow-chart representations of this are featured in Figure 48.

Can you draw the flow chart for a failure mode? Let us review some prevention mechanisms in more detail.

System redundancies and backups.

- Redundancies are system components designed to provide backup in critical situations—in case something goes wrong.
- Build redundancies and backups into the process; buy time until problem is resolved—e.g., emergency power. For example, uninterrupted power systems and emergency generators provide critical electric power for hospitals during regular power outages.

Failure modes. Often, it is impractical or uneconomical to design the system to withstand any imaginable catastrophe without failing. It is better to break up a single potentially catastrophic failure into a set of smaller, more manageable failures. A good example is the "crumple-zones" on the sheet metal of a car, which are designed to take the impact of a crash, deform, and divert damage from the occupants. Another is a reinforced concrete building, where the steel reinforcing bars are designed to fail before the concrete to avoid a sudden, "explosive" collapse, thus allowing the occupants some lead time to evacuate.

When we discussed crisis diagnostics in Part I, we were concerned with identifying *critical points* in the path of failure modes. Now, as a part of the system redesign, our objective is to take the critical points out of failure mode paths and replace them with *weak points*. In order to protect the most critical components, the system is designed to absorb the impact of a catastrophe and divert it from the most critical components and toward the least critical components. Thus, the least important parts are sacrificed to allow the most important to survive.

The least important element of a system that is designed to fail first is called the *weak point*, while the path of the impact through the system and to the weak points is called the *failure mode*. Complex systems can have multiple weak points and failure modes. By tracing the failure modes thorough the system map, we can predict the succession in which the weak points will fail and how the system will "look" after the impact. Computer simulations can predict the shape of the crumpled sheet metal after a car crash or how a building will look after an earthquake.

Practical risk management—ensuring the downside. A combination of the earlier-described methods is the best solution. The system must have "warning lights" to indicate that a failure has occurred and that the back-up systems are engaged. Implementation of crisis prevention mechanisms must be limited to (surprise!) crisis situations. For example, emergency generators in buildings don't kick in to provide extra electricity for Christmas lights; they engage in case of power failures to provide critical backup. Do not draw on that extra line of credit or force people into overtime unless conditions really warrant it. These are your reserves.

Remember: Always look around you for real-world examples to apply in your business management. Based on the earlier analysis of crisis-prevention methods in engineered systems, let us summarize them with business application examples. As an exercise, in Table 5, try to fill in the typical applications and examples of crisis-prevention mechanisms in your business.

Organizational Component—Finding the Right Way

Enforce Unity of Command and Action

Eliminate nonproductive and nonvalue-adding activities, procedures, and people. Always apply the lowest common denominator to skill/task matching: each activity should be performed by the person most qualified to perform it—the person for whom this is a high-end task. Do not waste highly skilled people on low-skill tasks. As people progress, promote them to their highest level of competence, but no higher; avoid promoting to their

Method	Application	Examples
System Redundancies		
Multiple Channel Processing		
System Backups		
Safety Factors		
Failure Modes		
Containment/Isolation/Quarantine		

Table 5. Crisis-Prevention Mechanisms

level of incompetence (Peter Principle). People must be competent and proficient in performing their tasks, but only just so—build stretch into the assignments; ensure that everyone is working at that fine line of challenge, between being bored and overwhelmed.

Implement Standard Operating Procedures

Establish a system of checks and balances—control and reporting mechanisms:

- ✔ Objective is clarity/transparency.
- ✔ Report exceptions to the rules, not the rules. In order to report exceptions from normal conditions, the range of "normalcy" must first be established. When does the red light go on?
- ✔ Reporting systems generate data—must be meaningful, accurate, timely, and regular.
- ✔ Good reporting systems are essential to develop a factual basis for decision-making.

Establish Clear Lines of Reporting and Authority

Everyone must know who's on first and where the buck stops. The "zero defects" culture of teams and shared responsibility is a dead-end road paved with good intentions. Mistakes and errors will continue to exist, regardless of whether we admit to them. Denial will not make them go away. Honest admittance and a willingness to address their causes will help reduce mistakes. Our pursuit of statistically perfect performance is meaningless and demeaning. Injured workers are sedated and brought back to work for light duty to artificially "avoid" lost-time accidents. Nobody is responsible for the injury since officially it did not happen. Fear of failure is paralyzing and breeds an irresponsible, a childish, and an artificially conformist culture or responsibility avoidance, forced consensus, and denial. Remember, if you work, you are in charge of your job and responsible for the results of your work. You must know who reports to you and to whom you report. Everybody has a boss and is a boss regardless of the pecking order. Even the CEOs have boards to whom they report, and the President of the United States can be voted out of office every four years. Somebody, somewhere must always be in charge of something.

Ensure That Authority Is Commensurate with Responsibility

People must be vested with equal levels of authority and responsibility to act. If someone is held responsible for accomplishing an objective, she must have the authority to do what is necessary to accomplish it. As we discussed in Part I, taking responsibility for a problem is the second step in the proactive crisis-resolution approach, after admitting that a problem exists. Subordinates cannot have more responsibility but less authority than the boss, an all too common situation in today's business organizations. Statements such as, "I don't want to hear about it," "Just tell me when you are done," and "I don't care how you do it; just get it done" are symptom code phrases of a manager trying to mask his incompetence by shifting his own responsibility to a subordinate. It is like being a combat soldier under fire in the trenches, yelling in vain, "Orders, Sir!" to your superior officer, who is in turn shell shocked, paralyzed with fear, and incapable of making a decision.

Provide Clear Task Definition

After all is said and done, there is always a lot more said than done.

Alfred Newman

Clearly define what is to be accomplished and by when—with firm completion deadlines.

Enforce Responsibility Assignment, Transfer, and Overlap

All tasks must be assigned to someone, and that someone has the responsibility for completing them. Eventually, tasks come to an end, or the responsible parties change. When does the responsibility for a task end? It is critical to ensure that each task is handed over and received "in good condition." All too often there are moments (or eons) of "weightlessness," when the responsibility belongs to no one—a responsibility gap. One party has terminated its responsibility and assumes that someone out there has taken over.

Few people remember the 1980s case of the Canadian 767 that ran out of fuel in midair and had to glide to an emergency landing, all because the required quantity of fuel (in gallons, I believe) was pumped in liters! At the refueling stop, the pilots noticed a low gauge reading, but it was not their responsibility to make sure that the tanks were full. The mechanic verified that the gauges were OK, but it was not his responsibility to make sure that the plane had enough fuel. The fueling crew members knew they had to pump 22,000 of something, which they logically assumed was liters, since Canada had just switched to the metric system. But it was not their responsibility to know where the plane was going next, to know the fuel consumption rate, or to worry about whether the fuel was sufficient to reach the destination. Thus, since nobody was responsible, nobody was at fault—ignorance was bliss, until the emergency landing. What happened here was a classic case of task compartmentalization and failed responsibility transfer. Although everyone was responsible for a facet of the overall task, nobody was in overall command. There was a major responsibility gap in the operating procedure. The best safeguard against improper responsibility transfers is to enforce responsibility overlaps. The transferor and the recipient work in parallel until the transferor is satisfied that the task is in good hands. This works only if both parties are cross-trained in each other's jobs—like a SWAT team.

Ensure Prompt Execution of Decisions—Certainty of Action

> As a confirmed melancholic, I can testify that the best and maybe only antidote for melancholia is action. However, like most melancholics, I also suffer from sloth.
>
> Edward Abbey

Along with assigning tasks, assign deadlines; a task without a deadline is not a task. Ensure continuity of effort and follow through until the final completion: start, sustain the effort, and finish every task.

How to stop procrastinating.

> The idle are a peculiar kind of dead who cannot be buried.
>
> Chinese Proverb

Recognize that procrastination stems from habit. New habits will be needed, and these take time and commitment to develop.

Procrastination results from:

- Real or apparently unpleasant tasks.
- Difficult and complex projects.
- Indecision.
- Fear of failure—If you are uncertain of your skills, starting a project will be tough and delaying will be easy. (Sometimes fear of success can be harmful as well.)
- A desire for perfectionism—You can't go on until you get it perfect.
- Lack of interest in the work.
- Hostility toward the person who assigned the project.

Human Component—Finding the Right People

> In the depths of my heart I can't help being convinced that my dear fellow men, with a few exceptions, are worthless.
>
> Sigmund Freud

> If you aren't fired with enthusiasm, you will be fired with enthusiasm.
>
> Vince Lombardi

What are the basic human qualities needed for the success of an organization—qualities that are rewarded by good business organizations?

Relevance of Knowledge, Skills, and Experience

Find out who are the best and smartest people in the company; if there aren't any, hire them, give them sufficient resources to do their job—and get out of their way.

Task/Skill Matching = Competence and Effectiveness

> No one is useless—even the useless can serve as a bad example.

Follow the Peter Principle—everyone is promoted to her level of incompetence. The role of a competent manager in the context of crisis avoidance is to ensure that everyone is competent at her job.

Applied Conventional Intelligence

> Perseverance is the most overrated of traits, if it is unaccompanied by talent; beating your head against the wall is more likely to produce a concussion in the head than a hole in the wall.
>
> Sydney Harris

Perhaps the most valuable commodity in crisis management is the competence and capability of the people involved. More than experience, knowledge, and expertise, the raw capacity to reason (conventional intelligence), coupled with practicality and effectiveness, is essential. Here are some traits of what I call *applied conventional intelligence*; they include the ability to:

- Reason/judge critically and logically.
- Think deductively—draw inferences.
- Think inductively—compile details to generate a larger picture.
- Extrapolate and project.
- Draw logical conclusions.

Nobel economics laureate Herbert Simon, in the third edition of his classic book *Administrative Behavior* (1976) discusses the human ability to rationalize. Here is my sumary of his two types of organizational personalities:

Economic man:

- Knows all alternative courses of action.
- Knows with certainty all consequences of those alternatives.
- Has a stable preference function for all consequences.
- Will maximize values by choosing the alternative followed by a preferred set of actions.

Administrative man:

- Has not the wits to maximize.
- Has bounded rationality imposed by this limited ability to process information.

Which one do you want to be?

Mental Fortitude—Moral Fiber

It's amazing how much you can accomplish if you don't mind who gets the credit.

Obsolescence—current managers are often less qualified and capable than their subordinates. Significance of experience (length of doing the same thing over and over again) is becoming drastically less important—what does twenty years of experience in the computer industry mean without continuous skill upgrades? What is needed is intelligence, flexibility, and courage to tackle problems. The incompetent will protect each other and themselves at the cost of the interests of the project or the organizations.

After reading the human component requirements, compare them to the human causes of crises identified in Part I; notice that they are exactly the opposite—good people will not let problems go unnoticed, fester into a crisis, avoid responsibility, and fail to actively resolve them. Conversely, a good organizational structure and system will make it impossible for the incompetent to thrive.

The following mechanisms will ensure that the organizational structure is not conducive to responsibility avoidance:

- ✔ Bias toward action.
- ✔ No heads in the sand/nothing falls between the cracks.
- ✔ Crisis situation gets identified and promptly addressed.
- ✔ Infertile ground for festering/smoldering of potential crisis situations.
- ✔ Prevents hiding behind teams, committees, groups—responsibility dispersal.

A committee is a group of people that can't do anything, who hold a meeting to determine that nothing can be done.

To reiterate, all three components of a successful business must be present: systems, organization, and people. Good systems and organizational structure are useless without good people to implement them; conversely, good people will leave a company with poor systems and organization.

It is true, however, that two good components, or even one good component, could cover and compensate for bad components for a while. For example, good people could still accomplish things in a poorly organized company with poorly designed systems, but they will expend unproductive energy to circumvent roadblocks and bypass red tape. Companies with rigid "fill in this form and stand in line" bureaucracies may have extensive procedures for handling trivia, but this should not be confused with efficient and effective systems and organizational structures that facilitate productive work. Complexity of systems and organizations is not related to productivity and the ability to prevent crises. As Thoreau put it, "any fool can make a rule." The other extreme is a totally lax and sloppy company, where systems and organization are absent and good employees waste time in creating and recreating logical work processes.

On the other hand, substandard people can survive for a while in well-organized companies, where the systems will push and prod the unwilling or unable to perform. But, this can only last for so long. No amount of stimulation will turn slackers into achievers, and they will eventually either be fired or voluntarily leave for—you guessed it—poorly managed companies, where their abilities and attitude are more in sync with the prevailing organizational structure and systems.

This matching of people to systems and to organizations is what is broadly called the "corporate culture," and mismatches between the components are main causes of failure to "fit in." Good people do in fact "fail" in bad companies. In other words, every company gets the employees it deserves (that fit in), and vice versa.

Following is a case that illustrates an attempt to address a crisis on a corporate level, by formulating a crisis recovery and prevention plan, which integrates the organizational, systems, and human components of the corporation.

A Case in Point—Back from the Brink

> We used to be on the brink of disaster ... now we're taking giant strides forward.
>
> Anonymous

This is a brief excerpt from a crisis recovery and prevention proposal submitted by the Project Surgeon to a major company, which was going through a severe transition period after being acquired by a competitor.

In light of the recent and impending resignations in your company, a number of the younger-generation managers have been voicing concern about the current status and future direction of Projectcorp Inc. This narrative is a synopsis of opinions from your veterans and newcomers alike, based on our informal conversations, and most importantly on the "debriefings" of people who have left. It identifies the biggest problems the company is facing today, and proposes specific and proactive measures for their resolution.

Major Problems

Resignations

Projectcorp is losing key young people to competition, and, along with them, the "goodwill assets" of the company, including invaluable skills, training, and talent. It is not the "dead wood" that is leaving, but the most marketable individuals. The result is a brain drain and the loss of the critical corporate knowledge base that is weakening your competitive advantage in the industry. Obviously, the best and the brightest would not have left if they weren't "ripe for the picking," i.e., if they had confidence and a stake in the future of the company. U, V, W, X, Y, and Z have all left, and we have not discovered or addressed the underlying reasons for their departure.

Undefined Long-Term Objectives and Direction

The overall opinion is that there is no strategic direction in the company, and that the employees are being "managed" on a day-to-day basis. What is your strategy in terms of market penetration, services, beating the competition, etc.? This uncertainty creates anxiety that is eroding employee morale and productivity. Furthermore, perception in the marketplace is that you are in a period of crisis and uncertainty, which further exacerbates internal problems.

Lack of Centralized Management and Decision-Making

There is no central operational authority that sets and implements corporate procedures, and provides overall guidance on a day-to-day basis. The line employees do not know where the "buck stops," your "standard operating procedures," or who sets the system of checks and balances that defines everyone's responsibility and authority.

Top-Heavy Management, Low Executive Productivity

The ratio of executives to project managers and engineers is close to 1:1, which means that for every fighting soldier in the trenches, there is a general in the headquarters. Your current market of smaller projects cannot support this structure.

Poor Housekeeping

Last but not least, overall apathy is manifested in abandoned workstations strewn with old drawings, files and boxes of projects long done, and lunches and naps on the digitizing table in the estimating room.

Recovery Objectives

To formulate a crisis recovery and prevention plan, we propose to establish a set of specific performance objectives, which will be implemented in a specific time frame and will each have a set of measurable performance indicators. It is important to focus on a few specific and achievable objectives whose accomplishment will pull your company out of trouble.

The following three guidelines should be taken into account to formulate your objectives correctly.

1. *Identify key areas in your business that are most in need of improvement*. You should objectively evaluate your strengths and weaknesses and develop a list of critical areas. Employee and client surveys will also help you acquire a balanced and unbiased perspective.
2. *Narrow the objectives to a manageable number, in order of priority and importance*. To maintain focus, concentrate on a limited number of areas highly important to your business performance. You cannot spread your efforts too thin. Performance objectives must not be too broad; otherwise you will not be able to address specific problem areas, and performance indicators will not be effectively measured.
3. *Determine indicators that are measurable, meaningful, and influenceable*. Performance in key areas must be measurable in comparison to set benchmarks within a specified time period. Once the benchmark is achieved, it should be raised to ensure that the performance improvement is continuous. The measurements must be meaningful and produce the desired improvement, i.e., you must measure the right things.

Following is a proposed list of major objectives and performance indicators.

Key Performance Areas and Indicators

1. **Marketing**

   Main Objectives:
   - Increase the number of projects won, from three this year to fifteen.
   - Increase annual volume from $50 million to $350 million.

   Performance Indicators:
   - Increase in jobs won as a percentage of proposals submitted, from 11 percent to 35 percent.
   - Decrease in marketing expense as a percentage of backlog, from 22 percent to 5 percent.

2. **Customer Satisfaction**

   Main Objective:
   - Ensure long-term partnering relationship with key clients.

   Performance Indicators:
   - Decrease in major customer complaints, from seventy-six per project to five (zero is the real target).
   - Increase in positive ratings in client feedback surveys.
   - Increase in repeat jobs from 33 percent to 75 percent.
   - Increase in long-term partnering or program management contracts.
   - Increase in outsourcing programs.
   - Increase in service diversification: facility management, etc.

3. **Employee Training and Development**

   Main Objective:
   - Ensure minimum turnover and continuous skill improvement of all employees.

   Performance Indicators:
   - Decrease in voluntary turnover, from 38 percent annually to 10 percent; increase in average length of employment, from three years to seven years.
   - Increase in skill enhancement: advanced degrees or registrations attained, special skills improvement.
   - Increase in computer software and MIS technological literacy.

4. **Project Operations**

   Main Objective:
   - Ensure that construction projects are completed under budget, on schedule, and within specified quality parameters.

   Performance Indicators:
   - Increase in bid/final cost underruns from negative 3 percent (loss) to positive 5 percent.

- Decrease in liquidated damages from $350,000 this year to zero.
- Decrease in warranty call-ups.
- Decrease in punch list and comeback work.

5. **Profitability**

Main Objective:
- Ensure profitability of business operations.

Performance Indicators:
- Increased net income per employee after taxes, from –$25,000 (loss) to $250,000.
- Increased return on capital employed.
- Reduced billing cycles to thirty days from current sixty-five-day average.
- Decreased overhead expense from 28 percent to 10 percent.
- Increased employee reimbursability from 60 percent to 95 percent.
- Increase ratio of revenue generating to support employees to 30:20.
- Delete nonproductive executive positions.

These are all achievable objectives and realistic target performance indicators. Successful accomplishment of these objectives will depend on the timely and rigorous implementation of the recovery plan.

Recovery Plan

Fill the Fighting Ranks with New Employees

Projectcorp must immediately hire a number of revenue-generating line employees to fill the void left by the recently departed people. No effort should be spared to hire and invest in the best people and demand best results in return.

Restore the Morale of the Remaining Troops

Projectcorp must rally the remaining line employees and convince them that they have a future with this company. A weekly project management meeting would be a good forum for the leadership to acknowledge the problems facing you, present a work-out plan, and ask for commitment and support. Also, clean up the office.

Establish Unity of Command and Leadership

Establish a COO position to centralize all procedural and operational functions. Delete redundant and overlapping levels of management. Restructure so that all line positions are revenue generating and report directly to the COO, with each position having clear definition of responsibility and authority.

Institute Mandatory Training Programs and Demand Skill Proficiency

With the recent resignations, critical technical knowledge is being irretrievably lost, e.g., MEP, structural, computers, which reflects negatively on the overall level of PMs' skills and knowledge. We must reestablish the mandatory training programs in critical skill areas, and demand that all new-generation PMs "graduate."

Implement standard operating procedures. Projectcorp must implement standardized "best practice" project management procedures, in order to eliminate redundant work. Each project manager now processes paper in a different way, none very efficient, and the wheel is reinvented on every new job. Rationalize and standardize the way you do work, from estimating, scheduling, and purchasing through document logs and closeout. Software exists that can be easily and inexpensively modified to the way we work. Our proposed Project Management Manual provides a primer for achieving this goal.

Implement stringent project performance controls and measures. Control, monitoring, and corrective procedures must be implemented to address the critical areas of project performance, including cost, schedule, and quality, on a regular (weekly or monthly) basis. For example, the monthly billing procedure should include job-cost-to-date status, profit/loss projection updates, receivables aging, etc. These reporting mechanisms should be summarized so that the COO knows the current status of all projects.

Once the major points of the Recovery Plan are in place, you can start focusing on achieving the specific performance objectives.

To summarize, here is a bullet-point summary of the overall assessment.

Problems:

- Unsuccessful penetration of new markets: services, project types and sizes, geographic locations.
- Erosion of existing markets and client bases—repeated losses to competition.
- Prolonged management transition—decision paralysis reflected on projects.
- Undefined long-term objectives/direction/role definition, locally and within corporate group.
- Potential financial exposure due to project losses.
- Top-heavy management, low executive productivity.
- Depletion of critical project management talent—brain drain, loss of corporate knowledge base.
- Low employee morale, erosion of accountability and commitment.
- Outside perception of crisis and uncertainty reinforces existing problems

Short-Term Recovery Measures—Up to Two Years:

- Restructure internally; remove bad apples and unproductive, overpaid, and underemployed people. Ensure that all remaining people earn their keep.
- Acknowledge problems and develop a work-out plan.
- Share plan with employees and ask for commitment and support.
- Maximize revenue by increasing immediate project volume—create financial breathing space.

Long-Term Recovery Measures:

- Define specific corporate role.
- Develop long-range marketing plan; follow through and monitor performance.
- Adapt to market; do not wait for the market to adapt to you.
- Invest in the best available project management technology.
- Invest in the best people and demand best results in return.
- Institute mandatory training programs and demand skill proficiency.
- Implement standardized "best practice" project management procedures,
- Define clear lines of accountability and responsibility.
- Define five to ten most important corporate performance areas and strive for continuous improvement.

The above program is just a first step in a long-term effort that must be sustained continuously in order to yield results.

Lessons Learned

What happened in the end? A number of positive changes have occurred, but the company never regained its old prominence. The competition is fast and ruthless, and lost market share was never fully recovered. However, the company has stabilized and found its niche market. The main lesson of this recovery plan was that, in a crisis, the causes inevitably point to the incumbent leadership. Guilty or not, the decline happened on the leaders' watch, and they do not want an outsider to place the blame at their doorstep. Therefore, in order to achieve a successful recovery, there must either be a change of top management, or the turnaround specialist must be empowered to implement the changes operationally. In this case, however, the executives of Projectcorp have discovered the enemy, and it was them ... and that was not the message they wanted to hear.

It is appropriate to close this book by reflecting on our own failings, since a continuing discrepancy between the required and actual competence to perform a job is the major cause of all business crises today. Following is a copy of my favorite "book"—if it can be called as such, because it is but a few pages long (in public domain). Elbert Hubbard wrote it in a single hour in 1899. Notwithstanding its somewhat archaic language and the long forgotten historical context of the Spanish-American war around which it was written, its also long-forgotten message of self-reliance, perseverance, and dedication to duty and the focus on getting things done is even more relevant now then it was back then. The central message of my book would be that to be a successful crisis manager, you must be able to "carry a message to Garcia."

A Message to Garcia

In all this Cuban Business there is one man stands out on the horizon of my memory like Mars at perihelion.

When war broke out between Spain and the United States, it was very necessary to communicate quickly with the leader of the Insurgents. Garcia was somewhere in the mountain fastness of Cuba—no one knew where. No mail or telegraph message could reach him. The President must secure his cooperation, and quickly.

What to do!

Someone said to the President, "There is a fellow by the name of Rowan will find Garcia for you, if anybody can."

Rowan was sent for and given a letter to be delivered to Garcia. How the "fellow by the name of Rowan" took the letter, sealed it in an oilskin pouch, strapped it over his heart, in four days landed by night off the coast of Cuba from an open boat, disappeared into the jungle, and in three weeks came out on the other side of the Island, having traversed a hostile country on foot, and delivered his letter to Garcia are things I have no special desire to tell in detail. The point that I wish to make is this: McKinley gave Rowan a letter to be delivered to Garcia; Rowan took the letter and did not ask, "Where is he at?"

By the Eternal! There is a man whose form should be cast in deathless bronze and the statue placed in every college of the land. It is not book-learning young men need, not instruction in this and that, but a stiffening of the vertebrae that will cause them to be loyal to a trust, to act promptly, concentrate their energies, do the thing: "Carry a message to Garcia."

General Garcia is dead now, but there are other Garcias. No man who has endeavored to carry out an enterprise where many hands are needed, but has been well-nigh appalled by the imbecility of the average man—the inability or unwillingness to concentrate on a thing and do it.

Slipshod assistance, foolish inattention, dowdy indifference, and half-hearted work seem the rule; and no man succeeds, unless by hook or crook or threat, he forces or bribes other men to assist him, or mayhap, God in His goodness performs a miracle, and sends him an Angel of Light for an assistant.

You, reader, put this matter to a test. You are sitting now in your office—six clerks are within call. Summon any one and make this request: "Please look in the encyclopedia and make a brief memorandum for me concerning the life of Corregio."

Will the clerk quietly say, "Yes, sir," and go do the task? On your life, he will not. He will look at you out of a fishy eye and ask one or more of the following questions:

- Who was he?
- Which encyclopedia?
- Where is the encyclopedia?
- Don't you mean Bismarck?
- What's the matter with Charlie doing it?
- Is he dead?
- Is there any hurry?
- Shan't I bring you the book and let you look it up yourself?
- What do you want to know for?

And I will lay you ten to one that after you have answered the questions and explained how to find the information and why you want it, the clerk will go off and get one of the other clerks to help him try to find "Garcia"—and then come back and tell you that there is no such man. Of course, I may lose my bet, but according to the law of averages, I will not.

Now, if you are wise, you will not bother to explain to your "assistant" that Corregio is indexed under the C's, not in the K's, but you will smile very sweetly and say, "Never mind," and then look it up yourself. And this incapacity for independent action, this moral stupidity, this infirmity of the will, this unwillingness to cheerfully catch hold and lift are the things that put pure socialism far into the future. If men will not act for themselves, what will they do when the benefit of their effort is for all?

A first mate with knotted club seems necessary; and the dread of getting "the bounce" Saturday night holds many a worker to his place. Advertise for a stenographer, and nine out of ten who apply can neither spell nor punctuate—and do not think it necessary.

Can such a one write a letter to Garcia?

"You see that bookkeeper," said the foreman to me in a large factory.

"Yes, what about him?"

"Well, he is a fine accountant, but if I'd send him uptown on an errand, he might accomplish the errand all right, and on the other hand, he might stop at four saloons on the way and when he got to Main Street would forget what he had been sent for."

Can such a man be entrusted to carry a message to Garcia?

We have recently been hearing much maudlin sympathy expressed for the "downtrodden denizens of the sweatshop" and the "homeless wanderer searching for honest employment," and with it all often go many hard words for the men in power.

Nothing is said about the employer who grows old before his time in a vain attempt to get frowsy ne'er-do-wells to do intelligent work, and her long, patient striving after "help" that does nothing but loaf when her back is turned. In every store and factory, there is a constant weeding process going on. The employer is constantly sending away "help" that has shown the incapacity to further the interests of the business, and others are being taken on.

No matter how good times are, this sorting continues; only, if times are hard and work is scarce, the sorting is done finer—but out and forever out, the incompetent and unworthy go. It is the survival of the fittest. Self-interest prompts every employer to keep the best—those who can carry a message to Garcia.

I know of one man of really brilliant parts who has not the ability to manage a business of his own and is absolutely worthless to anyone else, because he carries with him constantly the insane suspicion that his employer is oppressing, or intending to oppress, him. He cannot give orders, and he will not receive them. Should a message be given to him to take to Garcia, his answer would probably be, "Take it yourself!"

Tonight this man walks the streets looking for work, the wind whistling through his threadbare coat. No one who knows him dare employ him, for he is a regular firebrand of discontent. He is impervious to reason, and the only thing that can impress him is the toe of a thick-soled Number Nine boot.

Of course, I know that one so morally deformed is no less to be pitied than a physical cripple; but in our pitying, let us drop a tear, too, for the men who are striving to carry a great enterprise, whose working hours are not limited by the whistle, and whose hair is fast turning white through the struggle to hold in line dowdy indifference, slipshod imbecility, and the heartless ingratitude, which, but for their enterprise, would be both hungry and homeless.

Have I put the matter too strongly? Possibly I have; but when all the world has gone a-slumming, I wish to speak a word of sympathy for the man who succeeds—the man who, against great odds, has directed the efforts of others and having succeeded, finds there's nothing in it: nothing but bare board and clothes. I have carried a dinner pail and worked for a day's wages, and I have also been an employer of labor, and I know there is something to be said for both sides. There is no excellence, per se, in poverty; rags are no recommendation; and all employers are not rapacious and high-handed, any more than all poor men are virtuous. My heart goes out to the man who does his work when the "boss" is away, as well as when he is at home. And the man who, when given a letter to Garcia, quietly takes the missive, without asking any idiotic questions, and with no lurking intention of chucking it into the nearest sewer or of doing aught else but delivering it, never gets "laid off," nor has to go on a strike for higher wages. Civilization is one long, anxious search for just such individuals. Anything such a man asks shall be granted. He is wanted in every city, town, and village—in every office, shop, store, and factory. The world cries out for such: he is needed and needed badly—the man who can "Carry a message to Garcia."

Index

S

T

Upgrade Your Project Management Knowledge

with First-Class Publications from PMI

New Books

A Guide to the Project Management Body of Knowledge (PMBOK® Guide) – 2000 Edition

PMI's *PMBOK® Guide* has become *the* essential sourcebook for the project management profession and its de facto global standard, with over 700,000 copies in circulation worldwide. This new edition incorporates numerous recommendations and changes to the 1996 edition, including: progressive elaboration is given more emphasis; the role of the project office is acknowledged; the treatment of earned value is expanded in three chapters; the linkage between organizational strategy and project management is strengthened throughout; and the chapter on risk management has been rewritten with six processes instead of four. Newly added processes, tools, and techniques are aligned with the five project management processes and nine knowledge areas. For example, reserve time, variance analysis, and activity attributes are added to Chapter 6 (Project Time Management); estimating publications and earned value measurement are added to Chapter 7 (Project Cost Management); and project reports, project presentations, and project closure are added to Chapter 10 (Project Communications Management). This is one publication you'll want to have for quick reference both at work and at home.

ISBN: 1-880410-23-0 (paperback); ISBN: 1-880410-22-2 (hardcover); ISBN: 1-880410-25-7 (CD-ROM)

PMI Project Management Salary Survey – 2000 Edition

This 2000 Edition updates information first published in 1996 and expands coverage to over forty industry affiliations in nearly fifty countries in seven major geographic regions around the world. Its purpose is to establish normative compensation and benefits data for the project management profession on a global basis. The study provides salary, bonus/overtime, and deferred compensation information for specific job titles/positions within the project management profession. It also contains normative data for a comprehensive list of benefits and an array of other relevant parameters. *The PMI Project Management Salary Survey* – 2000 Edition is a vital new research tool for managers and HR professionals looking to retain or recruit employees, current members of the profession or those interested in joining it, researchers, and academics.

ISBN: 1-880410-26-5 (paperback)

Project Management for the Technical Professional

Michael Singer Dobson

Dobson, project management expert, popular seminar leader, and personality theorist, understands "promotion grief." He counsels those who prefer logical relationships to people skills and shows technical professionals how to successfully make the transition into management. This is a witty, supportive management primer for any "techie" invited to hop on the first rung of the corporate ladder. It includes self-assessment exercises; a skillful translation of general management theory and practice into tools, techniques, and systems that technical professionals will understand and accept; helpful "how to do it" sidebars; and action plans. It's also an insightful guide for those who manage technical professionals.

"The exercises and case studies featured here, along with the hands-on advice, hammer home fundamental principles. An intriguing complement to more traditional IT management guides, this is suitable for all libraries." —*Library Journal*

ISBN: 1-880410-76-1 (paperback)

The Project Surgeon: A Troubleshooter's Guide To Business Crisis Management

Boris Hornjak

A veteran of business recovery, project turnarounds and crisis prevention, Hornjak shares his "lessons learned" in this best practice primer for operational managers. He writes with a dual purpose—first for the practical manager thrust into a crisis situation with a mission to turn things around, make tough decisions under fire, address problems when they occur, and prevent them from happening again. Then his emphasis turns to crisis *prevention*, so you can free your best and brightest to focus on opportunities, instead of on troubleshooting problems, and ultimately break the failure/recovery cycle.

ISBN: 1-880410-75-3 (paperback)

Risk And Decision Analysis in Projects
Second Edition

John R. Schuyler

Schuyler, a consultant in project risk and economic decision analysis, helps project management professionals improve their decision-making skills and integrate them into daily problem solving. In this heavily illustrated second edition, he explains and demystifies key concepts and techniques, including expected value, optimal decision policy, decision trees, the value of information, Monte Carlo simulation, probabilistic techniques, modeling techniques, judgments and biases, utility and multi-criteria decisions, and stochastic variance.

ISBN: 1-880410-28-1 (paperback)

Earned Value Project Management
Second Edition

Quentin W. Fleming and Joel M. Koppelman

Now a classic treatment of the subject, this second edition updates this straightforward presentation of earned value as a useful method to measure actual project performance against planned costs and schedules throughout a project's life cycle. The authors describe the earned value concept in a simple manner so that it can be applied to any project, of any size, and in any industry. *Earned Value Project Management, Second Edition* may be the best-written, most easily understood project management book on the market today. Project managers will welcome this fresh translation of jargon into ordinary English. The authors have mastered a unique "early-warning" signal of impending cost problems in time for the project manager to react.

ISBN: 1880410-27-3 (paperback)

Project Management Experience and Knowledge Self-Assessment Manual

In 1999, PMI® completed a role delineation study for the Project Management Professional (PMP®) Certification Examination. A role delineation study identifies a profession's major performance domains (e.g., initiating the project or planning the project). It describes the tasks that are performed in each domain, and identifies the knowledge and skills that are required to complete the task. The role delineation task statements are presented in this manual in a format that enables you to assess how your project management experiences and training/education knowledge levels prepare you to complete each of the task statements. Individuals may use all of these tools to enhance understanding and application of PM knowledge to satisfy personal and professional career objectives. The self-assessment rating should not be used to predict, guarantee, or infer success or failure by individuals in their project management career, examinations, or related activities.

ISBN: 1-880410-24-9, (paperback)

Project Management Professional (PMP) Role Delineation Study

In 1999, PMI® completed a role delineation study for the Project Management Professional (PMP®) Certification Examination. In addition to being used to establish the test specifications for the examination, the study describes the tasks (competencies) PMPs perform and the project management knowledge and skills PMPs use to complete each task. Each of the study's tasks is linked to a performance domain (e.g., planning the project). Each task has three components to it: what the task is, why the task is performed, and how the task is completed. The *Role Delineation Study* is an excellent resource for educators, trainers, administrators, practitioners, and individuals interested in pursuing PMP certification.

ISBN: 1-880410-29-X, (paperback)

Other Bestsellers

PM 101 According to the Olde Curmudgeon

Francis M. Webster Jr.

Former editor-in-chief for PMI®, Francis M. Webster Jr. refers to himself as "the olde curmudgeon." The author, who has spent thirty years practicing, consulting on, writing about, and teaching project management, dispenses insider information to novice project managers with a friendly, arm-around-the-shoulder approach. He provides a history and description of all the components of modern project management; discusses the technical, administrative, and leadership skills needed by project managers; and details the basic knowledge and processes of project management, from scope management to work breakdown structure to project network diagrams. An excellent introduction for those interested in the profession themselves or in training others who are.

ISBN: 1-880410-55-9, (paperback)

The Project Sponsor Guide

Neil Love and Joan Brant-Love

This to-the-point and quick reading for today's busy executives and managers is a one-of-a-kind source that describes the unique and challenging support that executives and managers must provide to be effective sponsors of project teams. *The Project Sponsor Guide* is intended for executives and middle managers who will be, or are, sponsors of a project, particularly cross-functional projects. It is also helpful reading for facilitators and project leaders.

ISBN: 1-880410-15-X (paperback)

Don't Park Your Brain Outside: A Practical Guide to Improving Shareholder Value with SMART Management

Francis T. Hartman

Don't Park Your Brain Outside is the thinking person's guide to extraordinary project performance. Hartman has assembled a cohesive and balanced approach to highly effective project management. It is deceptively simple. Called SMART™, this new approach is Strategically Managed, Aligned, Regenerative, and Transitional. It is based on research and best practices, tempered by hard-won experience. SMART has saved significant time and money on the hundreds of large and small, simple and complex projects on which it has been tested. Are your projects SMART? Find out by reading this people-oriented project management book with an attitude!

ISBN: 1-880410-48-6 (hardcover)

The Enter*Prize* Organization: Organizing Software Projects for Accountability and Success

Neal Whitten

Neal Whitten is a twenty-three-year veteran of IBM and now president of his own consulting firm. Here he provides a practical guide to addressing a serious problem that has plagued the software industry since its beginning: how to effectively organize software projects to significantly increase their success rate. He proposes the "Enterprize Organization" as a model that takes advantage of the strengths of the functional organization, projectized organization, and matrix organization, while reducing or eliminating their weaknesses. The book collects the experiences and wisdom of thousands of people and hundreds of projects, and reduces *lessons learned* to a simple format that can be applied immediately to your projects.

ISBN: 1-880410-79-6 (paperback)

Teaming for Quality

H. David Shuster

Shuster believes most attempts at corporate cultural change die because people fail to realize how addicted they are to the way things are, the root causes of their resistance to change, and the degree to which their willingness to change depends on the moral philosophy of management. His new book offers a stimulating synthesis of classical philosophy, metaphysics, behavioral science, management theory and processes, and two decades of personal teaming experience to explain how individuals can choose change for themselves. Its philosophy-to-practice approach will help people team in ways that promote exceptionally high levels of bonding, individual creative expression (innovation), and collective agreement (consensus). Shuster shows how personal work fulfillment and corporate goals *can* work in alignment.

ISBN: 1-880410-63-X (paperback)

Project Management Software Survey

The PMI® *Project Management Software Survey* offers an efficient way to compare and contrast the capabilities of a wide variety of project management tools. More than two hundred software tools are listed with comprehensive information on systems features; how they perform time analysis, resource analysis, cost analysis, performance analysis, and cost reporting; and how they handle multiple projects, project tracking, charting, and much more. The survey is a valuable tool to help narrow the field when selecting the best project management tools.

ISBN: 1-880410-52-4 (paperback)
ISBN: 1-880410-59-1 (CD-ROM)

The Juggler's Guide to Managing Multiple Projects

Michael S. Dobson

This comprehensive book introduces and explains task-oriented, independent, and interdependent levels of project portfolios. It says that you must first have a strong foundation in time management and priority setting, then introduces the concept of Portfolio Management to timeline multiple projects, determine their resource requirements, and handle emergencies, putting you in charge for possibly the first time in your life!

ISBN: 1-880410-65-6 (paperback)

Recipes for Project Success

Al DeLucia and Jackie DeLucia

This book is destined to become "the" reference book for beginning project managers, particularly those who like to cook! Practical, logically developed project management concepts are offered in easily understood terms in a lighthearted manner. They are applied to the everyday task of cooking—from simple, single dishes, such as home-made tomato sauce for pasta, made from the bottom up, to increasingly complex dishes or meals for groups that in turn require an understanding of more complex project management terms and techniques. The transition between cooking and project management discussions is smooth, and tidbits of information provided with the recipes are interesting and humorous.

ISBN: 1-880410-58-3 (paperback)

Tools and Tips for Today's Project Manager

Ralph L. Kliem and Irwin S. Ludin

This guidebook is valuable for understanding project management and performing to quality standards. Includes project management concepts and terms—old and new—that are not only defined but also are explained in much greater detail than you would find in a typical glossary. Also included are tips on handling such seemingly simple everyday tasks as how to say "No" and how to avoid telephone tag. It's a reference you'll want to keep close at hand.

ISBN: 1-880410-61-3 (paperback)

The Future of Project Management

Developed by the 1998 PMI® Research Program Team and the futurist consultant firm of Coates and Jarratt, Inc., this guide to the future describes one hundred national and global trends and their implications for project management, both as a recognized profession and as a general management tool. It covers everything from knowbots, nanotechnology, and disintermediation to changing demography, information technology, social values, design, and markets.

ISBN: 1-880410-71-0 (paperback)

New Resources for PMP® Candidates

The following publications are resources that certification candidates can use to gain information on project management theory, principles, techniques, and procedures.

PMP Resource Package

Doing Business Internationally: The Guide to Cross-Cultural Success
by Terence Brake, Danielle Walker, and Thomas Walker

Earned Value Project Management, Second Edition
by Quentin W. Fleming and Joel M. Koppelman

Effective Project Management: How to Plan, Manage, and Deliver Projects on Time and Within Budget
by Robert K. Wysocki, et al.

A Guide to the Project Management Body of Knowledge (PMBOK® Guide)
by the PMI Standards Committee

Global Literacies: Lessons on Business Leadership and National Cultures
by Robert Rosen (Editor), Patricia Digh, and Carl Phillips

Human Resource Skills for the Project Manager
by Vijay K. Verma

The New Project Management
by J. Davidson Frame

Organizing Projects for Success
by Vijay K. Verma

Principles of Project Management
by John Adams, et al.

Project & Program Risk Management
by R. Max Wideman, Editor

Project Management Casebook
edited by David I. Cleland, et al.

Project Management Experience and Knowledge Self-Assessment Manual
by Project Management Institute

Project Management: A Managerial Approach, Fourth Edition
by Jack R. Meredith and Samuel J. Mantel Jr.

Project Management: A Systems Approach to Planning, Scheduling, and Controlling, Seventh Edition
by Harold Kerzner

A Guide to the Project Management Body of Knowledge (PMBOK® Guide) – 1996 Edition

The basic reference for everyone who works in project management. Serves as a tool for learning about the generally accepted knowledge and practices of the profession. As "management by projects" becomes more and more a recommended business practice worldwide, the *PMBOK® Guide* becomes an essential source of information that should be on every manager's bookshelf. The *PMBOK® Guide* is an official standards document of the Project Management Institute and will continue to serve as one of the reference documents for the Project Management Professional (PMP®) Certification Examination through 2001, after which the 2000 Edition will be used.

ISBN: 1-880410-12-5 (paperback), ISBN: 1-880410-13-3 (hardcover)

PMBOK Q&A

Use this handy pocket-sized, question-and-answer study guide to learn more about the key themes and concepts presented in PMI's international standard, *PMBOK® Guide.* More than 160 multiple-choice questions with answers (referenced to the *PMBOK® Guide*—1996 Edition) help you with the breadth of knowledge needed to understand key project management concepts.

ISBN: 1-880410-21-4 (paperback)

Also Available From PMI

Project Management for Managers
Mihály Görög, Nigel J. Smith
ISBN: 1-880410-54-0 (paperback)

Project Leadership: From Theory to Practice
Jeffery K. Pinto, Peg Thoms, Jeffrey Trailer, Todd Palmer, Michele Govekar
ISBN: 1-880410-10-9 (paperback)

Annotated Bibliography of Project and Team Management
David I. Cleland, Gary Rafe, Jeffrey Mosher
ISBN: 1-880410-47-8 (paperback) ISBN: 1-880410-57-5 (CD-ROM)

How to Turn Computer Problems into Competitive Advantage
Tom Ingram
ISBN: 1-880410-08-7 (paperback)

Achieving the Promise of Information Technology
Ralph B. Sackman
ISBN: 1-880410-03-6 (paperback)

Leadership Skills for Project Managers, Editors' Choice Series
Edited by Jeffrey K. Pinto, Jeffrey W. Trailer
ISBN: 1-880410-49-4 (paperback)

The Virtual Edge
Margery Mayer
ISBN: 1-880410-16-8 (paperback)

The ABCs of DPC
Edited by PMI's Design-Procurement-Construction Specific Interest Group
ISBN: 1-880410-07-9 (paperback)

Project Management Casebook
Edited by David I. Cleland, Karen M. Bursic, Richard Puerzer, A. Yaroslav Vlasak
ISBN: 1-880410-45-1 (paperback)

Project Management Casebook, Instructor's Manual
Edited by David I. Cleland, Karen M. Bursic, Richard Puerzer, A. Yaroslav Vlasak
ISBN: 1-880410-18-4 (paperback)

The PMI Book of Project Management Forms
ISBN: 1-880410-31-1 (paperback)
ISBN: 1-880410-50-8 (diskette)

Principles of Project Management
John Adams et al.
ISBN: 1-880410-30-3 (paperback)

Organizing Projects for Success
Human Aspects of Project Management Series, Volume One
Vijay K. Verma
ISBN: 1-880410-40-0 (paperback)

Human Resource Skills for the Project Manager
Human Aspects of Project Management Series, Volume Two
Vijay K. Verma
ISBN: 1-880410-41-9 (paperback)

Managing the Project Team
Human Aspects of Project Management Series, Volume Three
Vijay K. Verma
ISBN: 1-880410-42-7 (paperback)

Value Management Practice
Michel Thiry
ISBN: 1-880410-14-1 (paperback)

The World's Greatest Project
Russell W. Darnall
ISBN: 1-880410-46-X (paperback)

Power & Politics in Project Management
Jeffrey K. Pinto
ISBN: 1-880410-43-5 (paperback)

Best Practices of Project Management Groups in Large Functional Organizations
Frank Toney, Ray Powers
ISBN: 1-880410-05-2 (paperback)

Project Management in Russia
Vladimir I. Voropajev
ISBN: 1-880410-02-8 (paperback)

A Framework for Project and Program Management Integration
R. Max Wideman
ISBN: 1-880410-01-X (paperback)

Quality Management for Projects & Programs
Lewis R. Ireland
ISBN: 1-880410-11-7 (paperback)

Project & Program Risk Management
Edited by R. Max Wideman
ISBN: 1-880410-06-0 (paperback)

The PMI Project Management Fact Book
ISBN: 1-880410-62-1 (paperback)

A Framework for Project Management
ISBN: 1-880410-82-6, Facilitator's Manual Set (3-ring binder)
ISBN: 1-880410-80-X, Participants' Manual Set, (paperback)

Order online at www.pmibookstore.org

Book Ordering Information
Phone: +412.741.6206
Fax: +412.741.0609
Email: pmiorders@abdintl.com

Mail: PMI Publications Fulfillment Center
PO Box 1020
Sewickley, Pennsylvania 15143-1020 USA